Joost Beemer

Investing in Renewable Energy and Cleantech: a Finance Perspective

Joost Beemer

Investing in Renewable Energy and Cleantech: a Finance Perspective

an Academic View on the Returns of Renewable energy and Cleantech in Portfolio Management

VDM Verlag Dr. Müller

Impressum/Imprint (nur für Deutschland/ only for Germany)

Bibliografische Information der Deutschen Nationalbibliothek: Die Deutsche Nationalbibliothek verzeichnet diese Publikation in der Deutschen Nationalbibliografie; detaillierte bibliografische Daten sind im Internet über http://dnb.d-nb.de abrufbar.
 Alle in diesem Buch genannten Marken und Produktnamen unterliegen warenzeichen-, marken- oder patentrechtlichem Schutz bzw. sind Warenzeichen oder eingetragene Warenzeichen der jeweiligen Inhaber. Die Wiedergabe von Marken, Produktnamen, Gebrauchsnamen, Handelsnamen, Warenbezeichnungen u.s.w. in diesem Werk berechtigt auch ohne besondere Kennzeichnung nicht zu der Annahme, dass solche Namen im Sinne der Warenzeichen- und Markenschutzgesetzgebung als frei zu betrachten wären und daher von jedermann benutzt werden dürften.

Coverbild: www.purestockx.com

Verlag: VDM Verlag Dr. Müller Aktiengesellschaft & Co. KG
Dudweiler Landstr. 99, 66123 Saarbrücken, Deutschland
Telefon +49 681 9100-698, Telefax +49 681 9100-988, Email: info@vdm-verlag.de

Herstellung in Deutschland:
Schaltungsdienst Lange o.H.G., Berlin
Books on Demand GmbH, Norderstedt
Reha GmbH, Saarbrücken
Amazon Distribution GmbH, Leipzig
ISBN: 978-3-639-14120-7

Imprint (only for USA, GB)

Bibliographic information published by the Deutsche Nationalbibliothek: The Deutsche Nationalbibliothek lists this publication in the Deutsche Nationalbibliografie; detailed bibliographic data are available in the Internet at http://dnb.d-nb.de .
Any brand names and product names mentioned in this book are subject to trademark, brand or patent protection and are trademarks or registered trademarks of their respective holders. The use of brand names, product names, common names, trade names, product descriptions etc. even without a particular marking in this works is in no way to be construed to mean that such names may be regarded as unrestricted in respect of trademark and brand protection legislation and could thus be used by anyone.

Cover image: www.purestockx.com

Publisher:
VDM Verlag Dr. Müller Aktiengesellschaft & Co. KG
Dudweiler Landstr. 99, 66123 Saarbrücken, Germany
Phone +49 681 9100-698, Fax +49 681 9100-988, Email: info@vdm-publishing.com

Printed in the U.S.A.
Printed in the U.K. by (see last page)
ISBN: 978-3-639-14120-7

Investing in Renewable Energy and Cleantech:
a Finance Perspective

JOOST P. BEEMER[1]

Supervisor: Prof. dr. ir. H.A. Rijken

ABSTRACT

Renewable energy and cleantech, two emerging sectors are analysed extensively for the period 1994 to 2008 (ytd) with monthly data and compared with stocks, bonds and alternative asset classes. To study their ability to enhance the performance of an investment portfolio. Both sectors make a valuable contribution to an existing portfolio for investors preferring a portfolio with a higher return than 8.8 percent and who do not mind a standard deviation greater than 4.4 percent. When they do? They are better off with a portfolio excluding renewable energy and cleantech, holding only traditional assets.

Key words: Renewable Energy, Cleantech, Index Construction, Sharpe Ratio, Downside Risk Beta, Portfolio Choise, Asset allocation, Johanson Co-integration and Granger Causality.

[1] I thank Prof. Dr. Ir. H.A. Rijken for his excellent supervision and advice throughout the whole project. All remaining errors are my own •VU University Amsterdam • Faculty of Economics and Business Administration • De Boelelaan 1105, 1081HV Amsterdam, The Netherlands • info: www.feweb.vu.nl • contact : joost.beemer@kempen.nl • Thesis for the Degree of M.Sc. in Finance

In memory of my father

Contents

1.   Introduction

ULTIMATELY, ALL ENERGY PROVISIONS on Earth originate from the sun and solar energy provides a continuous flow of energy which warms us, causing biomass to grow via photosynthesis, heating land and sea, so causing wind power and consequently waves and rain leading to hydro power. Gravitational pull of the moon and the sun results in the rise and fall of water, creating tidal power. Geothermal heat is the result of radioactive decay deep in the Earth. As Dincer (2003) states, all are potential sources of energy, though the science is understood, it does not follow that: (1.) enough research money is poured into the project and (2.) an engineering solution should be found appropriately. The latter will be overcome by the former, when implemented effectively.

Paradoxically, while R&D investment has increased in almost all other industry sectors worldwide, it has steadily declined in the energy R&D sector since the 1970 global oil crisis, with the single exception of Japan (EIA, 2005). R&D expenditure for energy is quite dependant on the oil price. The oil price went down since the oil crises together with the R&D budget for energy. This relation is pointed out by Wustenhagen and Bilharz, (2006). R&D hitting its peak in the late seventies, with nuclear energy making up the biggest part of the energy R&D budget as it still does today (EIA, 2008). For a cleaner future, R&D needs to go up (Brown et al, 2001), although the renewable R&D part is rising within the energy R&D budget, it still has a long way to go and the economic benefits exceed the R&D costs by far (Dincer, 2003). Especially in U.S. there is a strong underinvestment in energy R&D. Already noticed in 1999, by Margolis and Kammen, the U.S. lag behind other countries' R&D budgets. They simply do not have to develop their level of awareness, with their large coal reserves and to a less extend their oil reserves, but more importantly, a present ruling energy (and foreign!) policy focused on preservation instead of innovation. Therefore it is good to see that other countries do significantly invest in renewable energy. Especially in Europe, where the majority of the funds invested in renewable energy flow to and where the most renewable start ups are located. In 2007 a total of US$ 148 billion was invested in renewable energy technologies, resulting in a total capacity of 241 GW of clean energy sources globally, excluding large hydro. Against 4300 GW total installed capacity, resulting in six percent generated renewable energy in 2007. When adding large hydro capacity to these numbers, the renewable energy share rises to 23 percent, against 18 percent in 2006 (Ren21 report).

This strong growth is needed too, because energy demand is estimated by the EIA (2005) to rise with 60 percent in 2030, i.e., 2 percent per year, because of global population increase and the increasing wealth world wide, especially in emerging markets. However, fossil based generated power still forms the majority of the energy consumed. To compound the political implications of this, most scientists now agree that the unusual changes in planetary climate are the direct result of greenhouse gasses added to the atmosphere, mainly from the combustion of fossil fuel. Hereby renewable energy can function as part of a solution to multiple social economic problems facing the world today: growing energy demand, too much emission, climate change and a high dependence on fossil based fuels. It also proves to be a tool to improve GDP as Chien and Hub show in their recent article in 2008. Renewable energy also has the capacity of changing vulnerable dependency of resources into autonomy, for instance, regarding the oppressive policy in Georgia and Ukraine by Russia.

As pointed out by Hartley (1990), renewable energy technologies represent a massive energy potential which dwarfs that of equivalent fossil resources. Politically this is recognized too, and in an increasing number of countries put into action, through changing the market conditions for renewable electricity producers by obligating utilities to buy renewable electricity and by dictating the price that renewable electricity producers would receive for their power. Normally these sorts of interventions should not be needed, but the pricing of the market does not incorporate future costs society will endure, when proceeding with the energy mix as it has been doing in the past decades. Therefore, it is necessary that regulations provide a stable investment environment for renewable energy (Wustenhagen and Bilharz, 2006, Sawin, 2001).

Renewable energy is a strong growth sector with high returns, a trend or sustainable, it provides much discussion on the internet, in the news and on television. In the media renewable energy is referred to as if it is an asset class, but is that the case? Does it hold distinctive characteristics to be of value to an investor? Renewable energy has got momentum and the need for renewable energy is clear, but is it a sound investment? There is no scientific literature that examines this issue from an investment perspective. Loads of literature exists, covering topics described above or about implementing renewable energy in the energy infrastructure (Wustenhagen and Bilharz, 2006, Painuly, 2001, Elliott, 2000, Haas, R. et al., 2004). However, there are no papers (let alone books) about the performance of renewable energy compared to traditional asset classes, like stocks or bonds. Or comparing it to generally accepted alternatives, like hedge funds, private equity, real estate, emerging markets or commodities. Though renewable energy is booming and investors step in en mass, no literature investigates if this is justified. Maybe because it is hard to get reliable data, the sector is relatively new. Nevertheless, an extensive analysis of renewable energy is possible, for example, with the use of an index, which represents the renewable energy sector. Therefore, renewable companies need to be identified. Renewable energy companies are companies that generate energy from renewable sources that are inexhaustible. Companies to be included in the index have to comply with the following criteria: (1.) At least 50 percent of the total power production of a company should be generated with renewable energy sources. (2.) At least 50 percent of the total sales of a company should comprise of the renewable power production as mentioned in criteria 1. The criteria are constructed in this way, because the kind of companies held by existing funds are often not renewable enough to represent the renewable energy sector. A company is renewable enough when more than half of its activities are renewable. For example, one owns a company when more than 50 percent of the shares are in possession. This same rational is applied to the inclusion criteria. The presence of non-renewable companies in the existing funds dilutes the pureness of the funds and consequently, the existing funds cannot provide a good quality analysis of the renewable energy sector.

There is nothing renewable about products that use energy more efficiently or products that lower emissions and consequently companies that produce them should not be included in the index. Renewable energy companies are also referred to as "pure play" companies. To keep the index as pure as possible will give investors a better chance for *direct-investing* in renewable energy and therefore enhances the possibility to conduct research about pure play renewable energy companies. There is no fund or index originated before 2008 that complies with the criteria, mentioned above. Companies often regarded as pure play do comply with criteria 1 but also pursue trading or consulting and advising activities that dilute their total renewable sales, making them more a service provider than a renewable pure play. To adhere to the renewable statement, those companies should never be included if the upmost purity is to be assured. An overview of the companies that comply with both criteria and are therefore included, is presented in Appendix III.

A new index is required to analyse renewable energy in its purest form, as currently no index is available that consists of pure play renewable energy producers only. Most indices labelled renewable consist of companies actually belonging to the cleantech sector, which consists of three subsectors; 1. Renewable energy, 2. Energy efficiency and 3. Decentralized energy. Only the first of the three subsectors represents the renewable energy sector. A non-pure play RE index blurs the analysis of the risk/return profile of renewable energy and its characterization within a portfolio. Therefore, a new Renewable Energy Index is created.

Often companies mentioned in the media and regarded by fund managers as 'renewable' generate much lower percentages of their production from renewable sources, but the companies are called 'renewable' anyway. Some examples of funds or indices that are not very specific in their definition of renewable are the ERIX (European Renewable Index), the EIC Renewable Energy Fund and the DNB Renewable Energy Fund. Unfortunately, funds that have similar holdings often use different names interchangeably. For example, naming their fund sustainable, green, alternative or clean. These names are probably marketing driven but are no good representation of the funds' holdings. To provide some clarity in the different names used, a brief explanation of the four different names is given in a way they should be used in recent literature. Although definitions are used, this thesis does not aspire to define the four mentioned labels, but an explanation is required to shed some light on the differences these labels hold. For instance, the formal definition of 'sustainable' mostly used is "development that meets the needs of the present without compromising the ability of future generations to meet their own need" (Anon., 1987). Thus, sustainable funds represent companies that pursue actions that can be maintained over a long period of time (actually indefinitely, but that is quite a statement). The second label 'Green' is a more popular word and therefore widely used and not one particular definition describes all its uses. However, in this case, Green funds constitute companies whose activities benefit or at least do not harm the environment. Alternative funds embody next to renewable stocks, stocks of companies that provide products which influence energy use, lower emissions, store energy for later use (decentralised energy) or energy efficiency. For example the New Alternative Fund (since 1982) constitutes cleantech holdings like Sharp and Philips. This illustrates the broad definition these funds represent and label the holdings as active in alternative energies. These stocks are more similar to technology stocks and they should be categorized under cleantech. For cleantech an existing index would be representative and the World Alternative Energy Index is chosen to approximate the performance of this sector. More elaboration will take place in the data description section.

The goal of this thesis is to shed light on whether it is wise to invest in renewable energy or cleantech. This study is relevant for portfolio managers, institutional investors, but also to investors in general and off course people interested in renewable energy. Purpose of this thesis is to inform investors if investing in renewable energy and/or cleantech is really different than investing in 'traditional' assets. The research question sought to be answered is the following: *why invest in renewable energy and cleantech, from a finance point of view*. The research is done by analysing the risk return profile, its correlations and performing a portfolio optimization to see if the efficient frontier increases when renewable energy is added to an investment portfolio. Using traditional tools like Markowitz portfolio theory, the Sharpe ratio and CAPM, but also using advanced co-movement measures like Johansen Co-integration and Granger causality. The contribution of this thesis extends in providing an extensive analysis of the returns on the renewable energy and cleantech sector, their relation with established asset classes and how both sectors perform together with those assets in a portfolio context.

A general excepted measure of risk and return is the Sharpe ratio (Sharpe, 1994). In this thesis the Sharpe ratio is used to compare the risk/return characteristics of the RE index with other asset classes. The risk/return profile is complemented with the market beta found using the CAPM (an extensive discussion concerning the CAPM is found in Fama and French (2003). Beta is the sensitivity of an asset's returns to changes in the market's returns. Modern portfolio theory states that the higher an asset's beta, the riskier is the asset. However, it is not entirely clear that investors think of (systematic) risk in this way. An asset can have a high beta if it tends to go up substantially more than the market when the market goes up, even if it does not tend to fall by more than the market when the market falls. In other words, a high-beta asset may magnify the market's upside swings and at the same time dampen the market's downside swings. Although such an asset would be considered risky in modern portfolio theory simply because of its high beta, most investors would probably disagree with this characterization. To measure this asymmetric relation the downside risk beta can be applied. Maybe less known, but downside risk has a long history in finance. Harry Markowitz, in his pioneering book Portfolio Selection already mentioned this notion. Ang et al (2004), Nawrocki (1999), Post and van Vliet (2004) and Estrada (2000) elaborate on this measure. The concept of downside risk has had a strong appeal to portfolio managers and investors. Pension fund managers in particular, given their underlying goal of protecting their principal and to minimize potential losses, depending on a target return, find downside risk

8

measurements quite useful. Investors associate risk with *bad* outcomes such as returns below their expectations. They do not associate risk with large positive returns, returns above their expectations. For this reason, investors' perception of risk is quite different than the modern portfolio theory definition of risk. Though standard deviation remains the usual way to report the risk of mutual and pension funds, the downside risk beta measure is added to the analysis because of its asymmetric risk measurement abilities.

After a risk/return profile for renewable energy and cleantech is formed, Markowitz portfolio theory is used to asses the performance of both sectors. Modern portfolio theory argues that the risk of an asset depends on the situation in which it is considered. When the asset is considered in isolation, its total risk is relevant. If it is part of a diversified portfolio, only its systematic (non-diversifiable) risk is relevant. Both measures are used in this thesis and renewable energy and cleantech, together with the other assets mentioned, are regarded as one portfolio with certain weights put on each asset. An efficient frontier is constructed of randomly modelling feasible portfolios. A frontier with renewable energy and cleantech and one without both sectors. In this way it is easily to see if the sectors contribute to an optimal asset allocation.

Asset allocation is important to investors and also used by pension funds. Bikker et al. (2007) studied stock market performance and pension fund investment policy and came to the following conclusions. (1.) Pension fund asset allocation is significantly driven by short term stock market performance. (2.) Pension funds do not automatically sell equities in rising markets but are more willing to buy equities after stock market corrections. (3.) Overall market timing does not add value (5 basispoints per year). The latter is particularly interesting, because pension funds do try to time the market. The first is interesting, due to the fact that renewable energy and cleantech achieved such high returns in the past three years, chances are high that pension funds would be seriously interested in both sectors. The second conclusion is particularly pleasant at the moment of writing, because due to the recent market turmoil, caused by the credit crisis and recession expectations, stocks of renewable energy and cleantech did not miss out on the drop in stocks prices all over the world. Prices are down 50 percent from levels less than six months ago, creating a perfect step in period for pension funds. The market capitalisation of the renewable energy sector is US$ 30 billion, that should be large and liquid enough for pension funds to initiate investments. Also after the recent distrust in complex derivatives, pension funds want to invest in transparent products and preferably with a mission related investment, due to their social function. Taking into account the fundamentals the sector will be back on its feet; demand for energy will rise, increasing back up by governments all over the globe (i.e., 20 percent in 2020 EU legislation), growing *green* awareness and strong technological improvements make demand for renewables continuous in the long term. Settled companies have pricing power; filled order books and little participants in the sector. Characterized by a high barrier of entry due to high technology factor, large investments required, permits are often needed, little supply of parts available and network connection is essential. This makes the prospects for renewable energy and cleantech promising to achieve stable cash flows in the coming years. And cleantech stocks that hold enough cash, so no borrowing is needed, should not have had those severe drops in prices as observed recently. However, this thesis is not a stock recommendation, but the rich investment environment need to be pointed out and thorough research is conducted to asses the performance of the renewable and cleantech sector. After the portfolio optimization this thesis follows an approach similar to the two-step estimator proposed by Engle and Granger (1987) to test for co-integration, but because of lack of relevance the tests are moved to the appendix. However, concerning these tests, first, the Augmented Dickey-Fuller unit root tests are carried out to examine whether the assets are first difference stationary series. Next, Johansen co-integration tests are performed for potential long term co-movements between the assets studied. Followed by Granger causality tests examining the short-term causal relations between the assets. The remainder of this thesis is outlined as follows: Section 2 describes the methodology used for constructing the index, the risk/return analysis and the portfolio optimization. Section 3 describes the data and in section 4 the results are presented and analysed. Ultimately, in section 5 a conclusion is given, suggestions for further research and conclusively an outlook about the future of renewable energy and cleantech is given.

2.   Methodology

This methodology section is divided into three subsections. The first subsection discusses the construction method for the Renewable Energy Index. Subsection B covers methods used for creating a risk/return profile for renewable energy. Subsection C describes the optimal asset allocation.

A.   Index Construction

The most general category of stock price indices is based on market capitalisation of companies. The full market capitalisation methodology takes the total number of shares issued by a company for computing the index into account. An other category is the free-float market capitalisation, which is a stock price index including only the free-float shares. Free-float shares are shares available at the stock market for trading. Thus, the market capitalisation of a closely held company would be reduced in a free-float based index, influencing the composition and eventually the return of the index.

Free-float market capitalisation for indices is considered to be superior to full market capitalisation because it better reflects the trading activity and liquidity in the market. The main reason behind the growing popularity of the free-float methodology, however, is that institutional investors generally track benchmark indices (Biswall, 2003). With free-float being an important factor due to the amounts possibly invested. However, the shortcoming of free-float market capitalisation is the transparency of *ownership data and democratization of the ownership,* which are key factors in implementing the free-float methodology. Comprehension about the size and horizon of certain block-holders, like management & employees, cross ownership, strategic holdings and government holdings are hard to determine. Professional index constructors come by this inadequacy by asking disclosure of the companies included and complement this with institutional filings (e.g., from the SEC). This demand lies, not surprisingly, beyond the abilities of the author and furthermore, in this study, the results will not change dramatically.

A critical point about the full market capitalisation weighted index method is that a low free-float lends itself to volatility and manipulation and rise in the price of the illiquid stocks (due to large holdings), which can cause the index to move sharply. Low free-float can also give rise to price manipulation where a group of investors or even a single speculator can move the prices leading to small investors end up paying a price, which is highly unsustainable (Biswal, 2003). Therefore, based only on free-float capitalisation, these indices could remove these anomalies to a large extent and reflect the underlying market movement in a better way if it is possible to ensure the above mentioned transparency of ownership data, which is not the case.

Datastream does provide a free-float factor, which can be used to alter the market capitalisation to a free-float market capitalisation based index. This alteration does not change the Renewable Energy Index significantly and no sharp moves were observed. However, also the Datastream free-float factor is not established without dispute and therefore in this thesis is chosen for a full market capitalisation based index.

Calculating the market capitalisation based index starts by calculating the returns and weights of individual companies included in the index. By multiplying its return by its respective weight, the contribution of a company is calculated. The index value is calculated by making a sum of all contributions of all companies in the index, multiplied by the index value of the previous trading day (GPR 250 Manual, 2008).

The returns for individual companies are calculated as follows:

$$R_{i,t+1} = \frac{(P_{i,t+1} + D_{i,t+1} - P_{i,t})}{P_{i,t}}$$

where

$R_{i,t+1}$ equals return of company i in period t, t+1

$P_{i,t+1}$ equals price of company i at time t+1

$D_{i,t+1}$ equals dividend on company i at time t+1

$P_{i,t}$ equals price of company i at time t

t equals the last trading day

The weights of individual companies in the index are derived as follows:

$$W_{i,t} = \frac{C_{i,t}}{\sum_{i=1}^{Nt} C_{i,t}}$$

where

$W_{i,t}$ equals weight of company i at time t

$C_{i,t}$ equals market capitalisation of company i at time t

N_t equals number of companies that meet the inclusion criteria at time t

An important note is that all market capitalisations of the companies used to calculate the weights must be denoted in the same currency. With the returns and the weights calculated, the index value is calculated as follows:

$$I_t = I_{t-1}\left(1 + \sum_{i=1}^{Nt} (W_{i,t} * R_{i,t})\right)$$

where

I_t equals index value at time t

B.   Risk/Return Profile

The Renewable Energy Index is constructed to measure the risk/return of renewable energy companies. A risk/return profile consists of the return, risk and correlation of the Renewable Energy Index and comparing them with benchmark indices. A general excepted measure of risk adjusted return is the Sharpe Ratio (Sharpe, 1994). In this thesis the Sharpe Ratio is used to compare the risk/return characteristics of the Renewable Energy Index with the market, with the alternatives, consisting of Hedge Funds, Private Equity, Real Estate, Emerging Markets and Commodities, and with substitutes of the Renewable Energy Index (other global alternative energy indices, although not pure play). Campbell (2004) concluded that extremely low (and negative) correlation with other asset classes results in a highly beneficial investment vehicle for an investors' portfolio. Therefore, the correlation of these categories with renewable energy and cleantech is analysed and a rolling correlation is performed to analyse the relation through time. The risk/return profile is complemented with the CAPM and the downside beta.

The Sharpe Ratio is a ratio developed by Nobel laureate William F. Sharpe[2] to measure risk-adjusted performance. The Sharpe Ratio is calculated by subtracting the risk-free rate, such as that of the three month U.S. T-Bill, from the expected rate of return for a portfolio and dividing the result by the standard deviation of the portfolio excess returns. The assumption is made that the historic returns represent the expected return of the next period. The Sharpe Ratio is stated as follows:

$$S = \frac{(\bar{R}_i - R_f)}{\sigma_e}$$

where

S	equals the Sharpe Ratio
R_i	equals return of a particular asset i
R_f	equals risk free rate
σ_e	equals standard deviation of the excess return (excess return is $(\bar{R}_i - R_f)$)

The Sharpe Ratio indicates whether a portfolio's returns are due to smart investment decisions or a result of excess risk. This measurement is simple and very useful because although one portfolio or fund can reap higher returns than its peers, it is only a good investment if those higher returns do not come with too much additional risk. The greater a portfolio's Sharpe Ratio, the better its risk-adjusted performance has been (Sharpe, 1994).

The Capital Asset Pricing Model (CAPM) was introduced by Sharpe (1964), Lintner (1965) and Treynor (unpublished). The CAPM is a model that describes the relationship between risk and expected return and is generally accepted and used in the pricing of risky securities. Although highly criticized and frequently proven wrong empirically, the model is industry standard, due to its straightforwardness and simple interpretation. The general idea behind CAPM is that investors need to be compensated in two ways: time value of money and risk (Bodie, 2002, Fama and French, 2003). The time value of money is represented by the risk-free (R_f) rate in the formula and compensates the investors for placing money in any investment over a period of time. The other half of the formula represents risk and calculates

---

[2] William Forsyth Sharpe won The Nobel Prize in Economics in 1990, "For their pioneering work in the theory of financial economics" (jointly with Harry Max Markowitz and Merton Howard Miller). www.nobel-prize.org

the amount of compensation the investor needs for taking on additional risk. This is calculated by taking a risk measure (beta) that compares the returns of the asset with the market over a period of time and to the market premium, the above mentioned excess return $(\bar{R}_i - R_f)$. The CAPM is defined as follows:

$$\bar{R}_i = R_f + \beta_i (\bar{R}_m - R_f)$$

where

R_m equals market return

β_i equals a risk measure beta

The CAPM says that the expected return of a security or a portfolio equals the rate on a risk-free security plus a risk premium. If this expected return does not meet or beat the required return, then the investment should not be undertaken.

The concept of downside risk is in a way similar to the CAPM, but if RE or cleantech is to provide an interesting investment alternative, it would be interesting to know how it performs when stocks perform below average. Ang et al (2006) reintroduce the concept of downside beta (β^-), which was first introduced by Bawa and Lindenberg (1977). The normal market beta applies to the following measure:

$$\beta = \frac{cov(R_i, R_m)}{var(R_m)}$$

In the case of downside beta, this equation changes to the following:

$$\beta^- = \frac{cov(R_i^e, R_M^e \mid R_M^e < \mu_m^e)}{var(R_M^e \mid R_M^e < \mu_m^e)}$$

where

R_i^e equals the excess return of a particular asset i

R_M^e equals the excess market return

μ_m^e equals the mean of the excess market return

And where $R_M^e| R_M^e < \mu_{r_i}^e$ stands for the market excess return given that the market excess return is below the average market excess return. This means the only observations taken into account when measuring the (downside) beta value, are the years in which the return is below the mean. The resulting beta value is called the downside beta. A negative downside beta indicates opposite movement to the market (Estrada, 2002). Put differently, when the market provides negative returns, this particular asset provides positive returns. These types of assets usually provide lower upside returns, but are still attractive to hold because of this feature (Ang et al (2006)).

Estrada sees the downside beta as looking at the semi variance of returns. And Estrada argues that downside beta is a more plausible measure of risk for several reasons: first, investors obviously do not dislike upside volatility; they only dislike downside volatility. Second, the downside beta is more useful than the normal beta when the underlying distribution of returns is asymmetric and just as useful when the underlying distribution is symmetric. In other words, the downside beta is at least as useful a measure of risk as the normal beta (Estrada, 2002). The downside risk beta is discussed in the empirical section together with the optimal asset allocation, because the downside characteristics of an asset are of importance to investors when they determine their allocation.

C.   Optimal Portfolios

For the portfolio construction mean variance is used, based on the modern portfolio theory founded by Markowitz (1952). The CAPM discussed above is derived from this theory. The principle that an investor chooses between two portfolios is the same as described earlier when choosing between two assets. The theory assumes that an investor will hold a portfolio with the lowest risk or when choosing between two portfolios with equal risk, the one with the highest return will be preferred. One can see, this is the same principle as the CAPM, additional risk is not necessarily appalling, if an investor is compensated for that extra risk with a higher expected return. An important conclusion Markowitz made was that an investor should not evaluate assets individually, but in a total investment portfolio. Therefore, it is important to look at the covariance structure of the assets within a portfolio. This is done with mean variance optimization. For a certain return a certain risk is minimized or vice versa. To add assets to the portfolio with low correlation with each other, portfolio risk can be partly diversified away. The systematic part of risk anyway. For the other part, asset specific risk, an investor should be compromised. Thus the correlations amongst the asset classes are of great importance and performance must be measured in a portfolio as a whole.

When calculating the optimal portfolios, they are formed on the basis of expectations. Here the fundamental assumption is made that expected returns, standard deviations and covariance structures are approximated by their historical values. To analyse if renewable energy and cleantech belong in an optimal asset allocation, portfolios are constructed. When calculating the optimal portfolios, renewable energy and/or cleantech should be included. An optimal portfolio is otherwise known as an efficient portfolio. Each portfolio constructed consists of all the asset classes included in this study. Every asset class is multiplied by a certain weighting, whereby all weights sum to one. This portfolio creation is done 5,000 times. By random numbering all the weights, 5,000 portfolios are constructed, which are all feasible by applying each weight to the corresponding asset within the portfolio. All feasible portfolios are plotted in a scatter diagram. The optimal portfolios are the portfolios that form the efficient frontier of all possible portfolios illustrated. The efficient portfolios lie on the envelope of all feasible portfolios, on the upper right edge of the scatter plot, with standard deviation on the horizontal axle and return on the vertical axle. In short, efficient portfolios are portfolios with the maximum portfolio return for a given standard deviation. Or vice versa, the minimum standard deviation given a certain return. An optimal portfolio is mathematically described as follows:

$$min \sum_i \sum_j x_i x_j \sigma_{ij} = var(R_p)$$

subject to

$$\sum_i x_i R_i = \mu = E(R_p)$$

$$\sum_i x_i = 1$$

where

x_i	equals the weight of a particular asset i
σ_{ij}	equals the covariance of two particular assets i and j
R_p	equals the return of a particular portfolio

The computations are performed in Microsoft Excel. The optimal portfolios, with the efficient frontiers, are displayed and discussed in the empirical section. Before the empirical section starts, the data are described extensively as well as more elaboration over the constituents of the indices, the Renewable Energy Index and the WAEX, representing the renewable energy and cleantech industry, respectively.

3.   Data Description

The Renewable Energy Index consists of companies that are pure-play renewable energy producing companies. The earliest share price data of a pure-play company is available on Datastream from January 1st 1987 onwards. Table 2 gives an overview of the number of companies included in the index through time. In 1990, share price data of two more pure-play companies became available. The renewable energy sector consists of six main technologies and in 1990 only electricity generated through hydro and biomass is represented by the three companies available. This does not give a good representation of the renewable sector and therefore, the index starts in 1994. By then, six companies are included, representing hydro, biomass and wind as generating technologies. From 1994 to 2008 nine more companies adhere to the criteria formulated earlier and now the renewable energy sector is represented by four electricity generating technologies including geothermal power.

Table 1
Number of companies included in the Renewable Energy Index through time

Year	1987	1990	1992	1996	2000	2008
Number of companies incl.	1	3	4	7	9	15

Solar and marine power lack representatives, because companies that generate power through these technologies do not primarily generate power or, unfortunately, do not hold enough renewable generation in their portfolio to be included in the index. Iberdrola Renovables[3] (Renovables is Spanish for renewable) included in the index after its initial public offering, on

---

[3] Iberdrola Renovables mainly an owner of wind-driven power plants, is traded in Madrid after its € 4.5 billion (US$ 6 billion) initial public offering priced at the low end of the expected range, selling 65% of about 845 million shares. It was the second-largest I.P.O. so far that year, behind the € 6 billion ($8 billion) offering of the VTB Group, a Russian bank, according to Dealogic (M&A information provider).

December 13th 2008. Iberdrola Renovables does generate solar power, but not significantly, to be observed in the numbers, as seen in Appendix III [4].

The World Alternative Energy Index, WAEX, is the index that best approximates the cleantech industry. WAEX includes the 20 largest stocks involved in renewable energy, energy efficiency (better use of energy generation, which involves industries such as energy meters and supraconductors) and decentralized energy supply (electricity generation in close proximity to the consumer, involving microturbines and fuel cells). WAEX is a market capitalisation weighted index, with a cap at ten percent so as to maintain an efficient diversification, meaning no stock can weight more than ten percent in the index. The WAEX is balanced every quarter and an index review takes place every six months (Society General website). As you see the WAEX completely complies to the given explanation of the cleantech industry earlier. The WAEX was launched on August 30th 2006. The data prior to the launch results from back testing. Back testing is calculating how the index might have performed prior to launch if it had existed using the same index methodology employed by Societé Generale today and based on the initial constituents of the Index.

Both the WAEX and the Renewable Energy Index are shown in Figure 1. The Renewable Energy Index rises from 100 to 523, after reaching its peak of 761 in July 2008. The WAEX rises from 100 to 1300, after reaching a peak of 1997 in the beginning of 2008. The indices are complemented with the MSCI World Index as a global stock market benchmark. All three indices are total return indices, which means price and dividends are included in the series displayed. The Renewable Energy Index shows hardly any increase before 2003, after a stable period of nine years a strong trend is observed till the second quarter of 2007. From Q2 2007 till beginning second half of 2008 a stable period is again observed, till the credit crises finally emerged and stock plumped globally, taking the renewable sector with its descend. A clear decreasing trend is therefore observed from Q3 2008 till today as the end of the crisis is still not in sight and recession has arrived in numerous countries. The renewable energy graph clearly shows the start of the boom in renewable energy in Q2 2003, fuelled by increased renewable favoured legislation throughout the world. According to the Worldwatch Institute[5],

---

[4] Appendix III exhibits the constituents of the Renewable Energy Index as of November 2008 with some illustrating facts concerning the technology weights of the index, the production and capacity of the companies included and exchanges traded, etc.
[5] An article by Worldwatch Institute, released on July 10th, 2003. *Renewable Energy Enters Boom Period.* Available on *http://www.worldwatch.org/node/1771*

the new Johannesburg Renewable Energy Coalition emerged in August 2007 out of the World Summit energy debate. The partnership consists of more than 80 nations that are committed to increasing their share of energy derived from renewable sources. The coalition is led by Europe, Latin America, and a group of small island states concerned about climate change. During this summit renewable energy was being recognized as a means to reduce the threat of

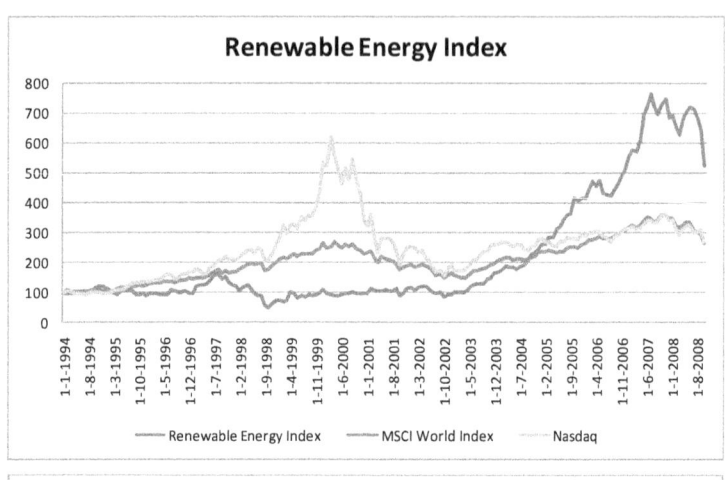

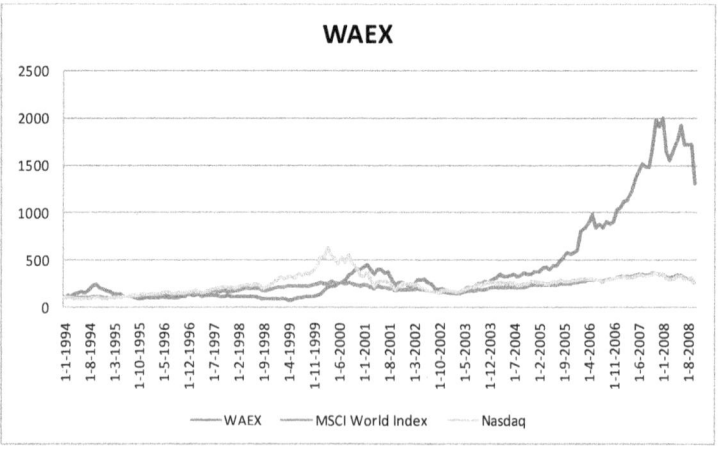

Figure 1: Descriptive Figures
The figures illustrate the Renewable Energy Index and the WAEX index through time ranging from January 1ˢᵗ 1994 till the end of October 2008, relative to the MSCI World Index represented as the global stock market and the Nasdaq representing the technology sector. All indices are total return indices, where January 1ˢᵗ,1994 is set as base date with value 100.

global climate change, to stimulate development, and to create jobs. Together with growing political attention in the United States[6], due to pressure since the U.S. are lagging behind the strong renewable growth in Europe. In Germany, by then the renewable leader in the world (EIA, 2005) the German government adopted its draft Renewables Act, which will reform existing legislation and redefine the framework conditions for renewable energy in the country (Wustenhagen and Bilharz, 2006). The initial proposal for this legislation was published in August 2003. The objective of the law is to increase the share of renewable energy in power production from 6.3 percent in 2000 to 12.5 percent in 2010 and to a minimum of 20 percent in 2020. This is in line with the EU target of increasing the share of renewable energy in terms of overall consumption from 6 to 12 percent (Wustenhagen and Bilharz (2006), EIA 2008, Ren21 report).

The same boom is visible in the WAEX graph, with the cleantech sector showing almost exponential growth, after 2003, with a short downfall in Q2 2006. Increasing even harder till the second half of 2007 and stabilising again till the end of 2007. In 2008, the WAEX descends as markets globally expect a recession and the credit crisis emerges. From 2003 to date the cleantech industry shows much resemblance with the renewable energy sector, a dip in 2006, a somewhat stable period in 2007 and a crash in 2008.

An interesting feature of the graph is the period surrounding the Dot.com bubble, from 1999 to 2001. On the contrary of what the stock market in general exhibits, does the renewable sector display a stable period, looking almost perfectly flat for three years. The cleantech sector shows a lagging peak in value. This indicates a lagging correlation with the technology sector for that period. A reason could be that investors where looking for alternative technology stocks after information technology stocks massively declined in value. After the Dot.com burst the cleantech sector shows no resemblance with the stock or technology market. The renewable sector shows no influence by the Dotcom bubble bursting. Indicating a lower correlation with the stock market and an even lower correlation with the Nasdaq. More elaboration on the correlations will be shown after the risk/return analysis.

---

[6] Increasing political attention, for example: In January, New York Republican Governor George Pataki announced an ambitious goal of generating 25 percent of New York's electricity with renewables by 2012. Labour unions demanding more renewable promotion in the form of the € 240 (US$ 300) billion 'Appolo project', increasing states issuing RPS acts (Renewable energy Portfolio Standards, meaning a certain percentage of renewable power production of total production) Worldwatch Institute (2003), EIA (2005).

The graphs show that the boom in cleantech was much higher than the boom in renewable energy. cleantech comprises of companies that deliver the technology to generate energy in a renewable way. cleantech funds invest in companies that deliver technologies as well as investing in companies that generate renewable energy, and therefore, they are better diversified than the Renewable Energy Index. Also, the stocks that were the stars of the last five years are mostly solar and wind stocks. Stocks that do not adhere to the criteria set earlier and are consequently represented in the cleantech index. It obviously shows that cleantech performed well due to that inclusion.

In this thesis monthly data are used. The data ranges from January 1st 1994 till November 1st 2008, except for the two substitute indices. For the Wilderhill and S&P Clean index the data ranges from January 1st 2001 and December 1st 2003 till now, respectively. It is important to note that fifteen years is a short range to perform the analysis conducted in this study. Other studies generally use 40 years or more to analyse performance of asset classes relative to other asset classes. Economic shocks, geopolitical events can have long lasting influences on the performance of certain assets, i.e. the Dot.com crises, war in Iraq and more recently the credit crises. Thus, reviewing the performance during a relatively short period of time, like fifteen years, can give a distorted view. Therefore, caution should be taken when interpreting the results. However, data about renewable energy and also for cleantech are not around so long and therefore the longest range possible is taken; 1. for the Renewable Energy Index to mean something and 2. to try to minimize the influence of extreme events.

Another critical point is important to emphasize: this thesis focuses on listed indices only and assumes that the displayed indices make a useful representation of the mentioned asset classes further on. Just as for private equity and hedge funds this holds for renewable energy, an investor could invest in non-listed funds or projects. For instance, setting up a windmill farm at a windy location or a PhotoVoltaic solar field in the Navarre desert in Spain (Faulina et al, 2006) could be a quite profitable investment. However, this kind of analysis complicates data gathering and lies therefore beyond the scope of this thesis.

All data in this study are downloaded from Thomson Financial Datastream, except the three month U.S. T-bill rate that was retrieved from FRED, Federal Reserve Economic Data in St. Louis, U.S. The data includes 178 observations. Table 2 presents the initial descriptive statistics for the data used. For the asset classes used in the study the indices that represent them are displayed, as well as the annualized mean return and standard deviation in percentages and the range of each asset.

Table 2
Descriptive Statistics

Table 2: displays descriptive statistics for the assets used in this study based on monthly data, the different asset classes will be used in the risk/return analysis, the portfolio optimization and the analysis of co-integration and Granger causality. Mean and standard deviation values are all in percentages and annualized ($\mu*12$ and $\sigma*\sqrt{12}$, respectively). The ranges are based on availability of a meaningful pure play Renewable Energy Index constructed especially for this study, which initiates in 1994. All other indices are total return indices obtained from Thomson Financial Datastream. The Three month U.S. T-bill rate is retrieved from FRED, Federal Reserve Economic Data in St. Louis, U.S.

Descriptive Statistics				
Asset	Represented by	Mean Return	Std. Deviation	Range
RE Index	Renewable Energy Index	15.29	28.07	1/1994 - 10/2008
Cleantech	World Alternative Energy Index	22.64	32.89	1/1994 - 10/2008
Wilderhill	Wilderhill New Energy Global I. Index	10.97	24.94	1/2001 - 10/2008
S&P Clean	S&P Global Clean Energy Index	19.15	25.39	12/2003 - 10/2008
Market				
Stocks	MSCI World Index	7.86	13.57	1/1994 - 10/2008
Corporate Bonds	Merril Lynch 5 - 7Y Corp. Bond Index	5.40	5.43	1/1997 - 10/2008
Government Bonds	J.P.Morgan Global Gov. Bond Index	0.37	3.26	1/1994 - 10/2008
Government 30Y Bonds	B. Deutschland 30Y Gov. Bond Index	3.69	12.51	1/1994 - 10/2008
Three Month U.S. T-Bill	Three Month U.S. T-Bill	3.88	0.49	1/1994 - 10/2008
Alternatives				
Hedge Funds	C.S/Tremont Hedge Fund Index	9.57	7.68	1/1994 - 10/2008
Private Equity	Liquid Private Equity 50 Index	9.83	21.71	1/1994 - 10/2008
Real Estate	World REITs Index	12.83	15.77	1/1994 - 10/2008
Emerging Markets	Dow Jones Emerging Markets Index	8.01	24.11	1/1994 - 10/2008
Commodities	Goldman Sachs Commodity Index	10.55	20.77	1/1994 - 10/2008

The stock market is represented by the MSCI World Index (Morgan Stanley Capital Index). The corporate bond market is represented by the Merril Lynch 5 to 7 years Corporate Bond Index and is chosen because it matches the average maturity of the J.P. Morgan Government Bond Index, which represents the government bond market. This study is conducted primarily for the group of investors who are interested in the risk and return profile of renewable energy and cleantech. This group includes institutional investors; for them the relation with long term bonds (30 year maturity) is of great importance. Since pension funds use long maturity bonds to match their long term (nominal) liabilities. The German bond market is very liquid and of substantial size to represent the global market for bonds with very long maturities. Therefore, the German 30 year government bond index is included in the analysis and will be elaborated upon in the discussion concerning the correlations in the empirical section. To calculate the excess returns and for the short term debt analysis the three month U.S. T-bill is used as an approximation for the risk free rate. Furthermore, the renewable energy and the cleantech sector are compared with generally accepted alternative asset classes, like; hedge funds, private equity, real estate, emerging markets and commodities. The Tremont Hedge Fund Index represents hedge funds as an asset class. This equally-weighted net-of-fees return index is broadly diversified across different style sectors. It could give more information to analyse the performance of the renewable energy- and the cleantech index relative to various categories of hedge funds. Because each hedge fund category exhibits different risk and return properties, especially varying with each categories horizon. More on these differences is elaborated upon by Ackermann et al. (1999) and Fung and Hsieh (1997). Rather than accounting for this heterogeneity in investment styles, the focus lies on an index which represents the whole industry, because in this thesis hedge funds are addressed as an asset class. In this same way private equity is represented as an asset class by the LPX 50, the Liquid Private equity Fund covering listed private equity funds and companies. The World REIT Index (Real Estate Investment Trust) is a leading global property index and therefore represents real estate as an asset class in the analysis. Emerging markets are represented by the Dow Jones Emerging Market Index. The GSCI, Goldman Sachs Commodity Index represents the last alternative asset category, commodities. It is a composite index of all world-production weighted commodity sector returns. This index represents an unleveraged, long-only investment in fully collateralized nearby commodity futures with full reinvestment. All indices are total return indices that include both price and dividend returns.

The Renewable Energy Index is further compared to indices that are liquid, with global constituents and most resemble the Renewable Energy Index. Two comparable indices are chosen. First, the Wilderhill New Energy Global Innovation Index (Wilderhill); it holds beyond renewable energy companies, cleantech companies that are involved in hydrogen and fuel cells, power storage and energy efficiency. Wilderhill holds three pure play renewable energy companies. Of the three indices, (including the cleantech index) the Wilderhill holds the most genuine renewable energy companies and is therefore chosen because it resembles the Renewable Energy Index the most. The second substitute index, the Standard & Poors Global Clean Energy Index (S&P Clean), which holds two pure play companies, resembles more the WAEX, World Alternative Energy Index. The WAEX in this thesis representing the cleantech sector holds no companies that comply to the RE criteria mentioned earlier. S&P Clean and the WAEX both have holdings that are active in energy producing and energy technology & equipment providing. Also all these substitute indices are total return indices.

If the renewable energy and the cleantech sector happen to correlate high with the stock market it can be interesting to analyse the correlation of renewable energy and the cleantech sector with each sub sector of the MSCI that form the MSCI World Index together. This creates the possibility to determine which sector relates best to the characterization of the renewable energy and cleantech sector. Maybe this information is of use to investors. Now that the data are described extensively, it is time to analyse the performance through the empirical findings, which will be analysed in the next section.

4. Empirical Analysis

The next step of this study is to compare the risk/return profile of renewable energy and cleantech with the traditional financial assets and apply Markowitz (1952) portfolio theory to the incorporation of renewable energy and cleantech in an existing mean-variance portfolio without the two assets, in order to check whether renewable energy and cleantech could be regarded as a useful portfolio component. The empirical analysis consists of the following subsections. First the risk/return analysis; in the second subsection the correlations are discussed and the last section is about portfolio optimization together with the downside risk comparison. The Johanson co-integration and Granger causality test results are discussed in the Appendix IV.

A. Risk / Return Analysis

To compare the performance of renewable energy and cleantech with other asset classes, the return statistics are displayed in Table 3. The Renewable Energy Index is self-constructed and the cleantech sector is represented by the WAEX. All observations are based on end period values. Note that for the two substitute asset classes, Wilderhill and S&P Clean, fewer observations were available. Therefore, these indices are not always included in the analysis, for example, in the correlation matrix. The three year rolling correlation through time does show the correlation of the substitute indices. Not all returns will be discussed, only the relevant statistics will be named. For a better interpretation of the figures, the mean, standard deviation and Sharpe Ratio are annualized and exhibited in Table 3. The annualized statistics are used when referring to in the analysis.

The returns of renewable energy and cleantech clearly outperform the traditional asset classes for the period 1994 – 2008 (ytd). Although measured during a shorter period, the substitute indices show returns in line with the returns of renewable energy and cleantech. An investment in stocks would have made an eight percent return in 15 years, compared to a 15 percent return in renewable energy and even 22 percent for WAEX on an annually basis. These returns are extremely high and this is probably due to the boom in green awareness as discussed earlier, together with an absence of profitable investment alternatives. After the internet bubble, listed traditional asset classes had a hard time. Before the bubble, simply too much money was around, chasing too few deals as pointed out by Gompers and Lerner (2000). The market overheated and asset classes like general stocks and listed private equity fell sharply and continued to fall till 2003. In the mean time there were alternatives, like hedge funds and real estate, performing really well during those years (as can be seen in figure 2), investors also fled to commodities and emerging markets. But there still was a lot of money to be allocated, especially seeking technology exposure. The IT-sector was less interesting, so another technology-driven sector was sought and found, renewable energy and cleantech.

When looking at performance, returns only give a one-sided perspective. Returns have to be reviewed in relation to the risk one endures while acquiring such return. If an investor takes on more risk than the risk of a benchmark asset, an investor should be compromised for taking on that additional risk.

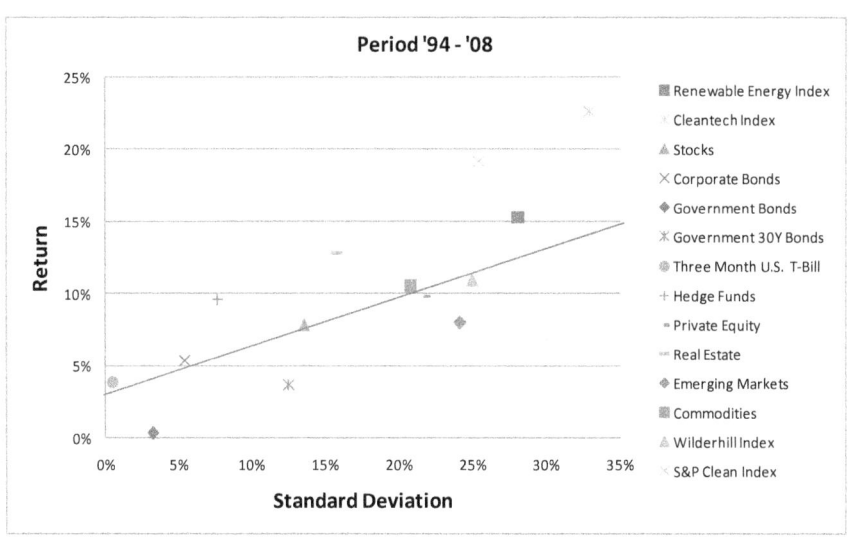

Figure 2: The Risk / Return of the Assets
The figure displays the risk relative to its return for each asset used in this study based on monthly data. Mean and standard deviation values are all in percentages and annualized ($\mu*12$ and $\sigma*\sqrt{12}$, respectively). The ranges are based on availability of a pure play renewable energy index constructed especially for this study, which initiates in 1994. All other indices are total return indices obtained from Thomson Financial Datastream. The Three month U.S. T-bill rate is retrieved from FRED, Federal Reserve Economic Data in St. Louis, U.S.

To verify if this is the case, the returns are compared relative to their volatility, displayed as the standard deviation. Figure 2 exhibits this relation, the red diagonal represents the benchmark. Starting from the risk free rate through the general stock market, this represents what investors could have earned, when simply investing in a proportion of these two assets. All assets above the red line have performed relatively well. The relation between return and risk is also represented in one number, the Sharpe Ratio, where returns are actually excess returns taken over the risk free rate (Three Month U.S. T-bill). The Renewable Energy Index shows a Sharpe of 0.41. This is much higher than the Sharpe of stocks, 0.29. So when taking risk into account, renewable energy would still be a better investment over the last 15 years than stocks. An investment in cleantech would have even resulted in a higher Sharpe of 0.57. Indicating cleantech as a superior investment to renewable energy. This is easily seen in figure 2: both are above the red line and cleantech lies above renewable energy. The higher Sharpe for cleantech resulted through a higher return while taking on just slightly more risk. The higher return is achieved through renewable energy technology (RET) stocks that went through the roof when policy makers recognized that it was inevitable for renewable energy to be part of a solution for the recent social economic problems if regulations were not there to support it (Wustenhagen and Bilharz, 2006). This en mass recognition started a rise in RET-stock prices from 2003 onwards. However, RET stocks are a kind of technology stocks and so cleantech stocks are more volatile than renewable energy stocks. Because renewable energy companies function more like utility companies; once the investment is made and capital is in place and running, renewable power is sold through PPA's (Power Purchasing Agreements), resulting in stable cashflows. Together with country regulations promising a certain price for renewable power, a stable investment environment is created. This should result in a lower risk for stocks of a renewable energy company. However, this is not displayed in the indices standard deviation, cleantech has almost the same volatility as renewable energy, 33 percent and 28 percent, respectively. This is maybe due to certain factors. which will be elaborated upon. The first explanation that renewable energy does not have a lower standard deviation than cleantech is perhaps that the cleantech index is better diversified. Cleantech includes more subsectors; renewable energy, efficient energy and decentralised energy. Furthermore it holds more different companies, therefore, it is possible better diversified than renewable energy. The second factor could be that investors perceive renewable energy companies as more risky, although future cash flows are in a way insured by regulation. Maybe because the technologies used are not fully proven yet. Or investors believe regulation may change, which endangers future income. A third reason that cleantech does not have a significant higher standard deviation may be due to the way the Renewable Energy Index is constructed. The renewable index consists of 15 companies (against 20 in the WAEX) till now. In the near future, more companies will be included and the volatility of one company will have a smaller impact on the index performance. Also the constituents of the Renewable Energy Index are the same during the whole period. Most indices include only the most liquid stocks and for that reason the composition changes, perhaps influencing the index volatility during a certain period. The latter reason may also be one of the reasons why the WAEX has such a high performance for such a long time, but this would also count for other liquid indices, i.e. LPX 50 (private equity).

But is the performance of renewable energy and cleantech that good? When comparing it to generally accepted alternative asset classes renewable energy and cleantech both are surpassed by hedge fund and real estate performance, with Sharpe Ratios of 0.75 and 0.57, respectively. Real estate is outperforming renewable energy and matching cleantechs performance. Hedge funds dominate all asset classes in terms of relative performance, although absolute return is not so high with 10 percent, it is achieved while taking on little risk, only 8 percent standard deviation in the returns, resulting in such a high Sharpe Ratio.

When looking at the substitute indices you see that Sharpe ratios are robust. Renewable energy, as mentioned, best in line with the constituents of the Wilderhill index (0.33), show similar Sharpe Ratios. The same accounts for cleantechs substitute the S&P Clean energy index (0.63). S&P Clean has a higher Sharpe, but data are available since 12/2003, right after the start of the boom, so caution is needed here and not too much weight can be put on this result. However, the corresponding Sharpe Ratios indicate the performance is representative for both sectors.

When looking at the Sharpe Ratios of the alternatives in Table . the private equity and emerging markets underperform stocks. Listed private equity took on some hard blows after an extreme peak during the top of the Dot.com bubble. Never reaching those levels again, ending up with a return of ten percent and a volatility of 22 percent for the last 15 years. Literature about private equity returns mention roughly five to ten percent in excess of the S&P 500 return (e.g. Ljungqvist and Richardson (2003), Cochrane (2005), Gompers and Lerner (2000)). Definitely listed private equity underperforms real private equity. Emerging market returns are also outperformed by the global stock market. Going down after 1997 (Asia crisis), only rising again after 2003 when BRIC stocks finally delivered. Government bond performance falls behind that of all other asset classes. Returns are low and relative to a riskless asset they become even negative, resulting in negative Sharpe Ratios. Investments in government bonds (both maturities) would have destroyed value if only held for the studied period of time. Thus the reason why these assets are held by investors must be other than risk/return only.

B.   Correlation and Downside Risk Beta

For an investor, the choice of including a certain asset class in its portfolio is not only about its risk/return performance, but about how that performance behaves in relation with the assets already in there. A measure for that interrelation is the correlation coefficient. Table 3 shows the correlations of renewable energy and cleantech with other asset classes. The correlations will be discussed together with the CAPM results and the downside risk beta noted in Table . The market betas of renewable energy and cleantech are insignificant, the factor loading is of importance here.

Renewable energy has a correlation of 0.54 with stocks. That is not very high; it indicates renewable energy is not as good as, for example, government bonds in diversifying risk in a portfolio. As indicated earlier, one of the reasons why government bonds are held by investors, is because of their low correlation with stocks. Other studies acknowledge this (Cambell and Viceira, 2000, Hoevenaars et. al, 2007). The market beta of renewable energy is more than one this indicates that movements of the general stock market are followed by larger movements of renewable energy stocks. The dowside risk beta of 1.19 indicates that a downward movement of the market on average is followed by a downward movement 1.19 times stronger than that of the market. This co-movement also indicates that renewable energy is not really a good diversifier for stocks. An interesting correlation is that of renewable energy and real estate. Real estate is considered a good alternative asset class for its distinctive return characteristics (Hoevenaars et. al, 2007). It is therefore held in many portfolios, a market beta of 0.5 affirms this relation. Just as the downside risk beta, confirming the reputable diversification characteristic of real estate, with a value of 0.5. Meaning a downfall of the markets return is followed by only half the drop in real estate return. The two have a correlation of 0.27; that's quite low and can mean that renewable energy is a good diversifier with respect to the other alternative asset classes. Table 3 shows that indeed the correlations with other assets are low, except with emerging markets. This can mean that in a portfolio, renewable energy could be included because of a valuable combination of characteristics; the good risk/return and the correlation with other alternatives. An answer to this question will be given in the next section, the portfolio optimization.

The same reasoning applies to the cleantech sector; it has a slightly lower correlation of 0.47 with stocks compared to renewable energy. And almost the same correlations with the alternative asset classes, except for emerging markets, which correlates less with cleantech as with renewable energy. The market beta of cleantech is higher than one and equal to renewable energy. The downside risk beta is also the same, 1.17. Thus a possible inclusion of cleantech in an optimized portfolio is not so much due to its correlation with stocks, but because of its interrelation with the other asset classes.

The notion was already raised in the methodology section, that next to the general stock market, also the correlation with the 10 subsections of the MSCI World Index is analysed for the renewable energy and cleantech sector. The results are shown in Appendix II. No high correlations were found. The highest were with the subsector Financials, but that holds no plausible explanation. The second highest is with Industrials, a correlation of for renewable energy 0.54 and cleantech 0.52. This is not a higher correlation than with the general stock market. To be conclusive, the correlations between the sub-subsectors of the MSCI Industrials and the Renewable Energy Index and the cleantech index were computed as well and again no higher correlation than with the general stock market were found. The highest correlation is with the Capital Goods Index of 0.53, equal for both sectors. Surprisingly, the correlations found for the MSCI Utility Index and the MSCI Energy Index were both lower than for general stocks. Thus, no interesting high correlating relations were found via the subsectors of the MSCI.

When looking at interesting correlations other than the low correlations of bonds mentioned earlier, commodities stand out. All correlations of commodities are low and a downside risk beta of 0.22, which is very low and, therefore, commodities are a great diversifier and safe haven when stock markets fall and great value in a portfolio. This is in line with the literature in which similar results are found (Brennan et al, 1997, Cambell and Viceira, 2000, Hoevenaars et. al, 2007). Other assets with low downside risk betas are hedge funds and bonds. Hedge funds per definition have low market betas, which corresponds with a 0.3 for both betas. Bonds are a substitute for stocks, therefore, they should have low betas, which is the case. From this reasoning follows that they should hold negative downside betas. Only both government bonds show a negative downside risk beta. Corporate bonds show a very small downside risk beta of 0.07. The positive signal is logical, because the stock market falls due the same influences that affect corporate bonds.

In a portfolio perspective it is important how the alternatives behave relative to each other, but because general stocks are an important benchmark in a portfolio. The correlation over time between the different asset classes and stocks is emphasized. A three year rolling correlation is estimated and presented in Figure 3. Figure 3a exhibits all asset classes and figure 3b shows only the rolling correlations of stocks with renewable energy and stocks with cleantech.

Table 3
Return Statistics

Table 3: displays the return statistics of all the selected asset classes. Panel A exhibits general descriptive statistics for the monthly data used, though extended with the annualized mean ($\mu*12$), the annualized standard deviation ($\sigma*\sqrt{12}$) and the annualized Sharpe Ratio, for a more easy interpretation. Panel B shows the results of the Capital Asset Pricing Model (CAPM). The return statistics are based on all monthly observations that were available between 1994 and 2008 ytd. All returns, medians, maximums, minimums and standard deviations are in percentages. CAPM estimations are based on excess returns, $Ri - rf = \beta i (Rm - rf)$, where the market return is represented by the global stock market (MSCI World Index) and the risk free rate is represented by the three Month U.S. T-Bill. All returns here are simple returns. See Table 2, Data description for more information on the different asset classes. Panel B shows the probability values in Italics.

Panel A - Return Statistics

Monthly Statistic	Renewable Energy	Cleantech	Stocks	Corporate Bonds	Government Bonds	Government 30Y Bonds	3 Month U.S. T Bill	Hedge Funds	Private Equity	Real Estate	Emerging Markets	Commo-dities	Wilderhill	S&P Clean
Mean	1.27	1.89	0.66	0.47	0.03	0.31	0.32	0.80	0.82	1.07	0.67	0.88	0.91	1.00
Median	0.82	2.77	1.24	0.60	0.11	0.11	0.38	0.80	0.78	1.36	0.67	0.89	2.69	0.00
Maximum	36.12	33.26	9.95	3.76	3.21	10.56	0.53	8.53	31.69	17.41	24.40	15.79	21.26	20.19
Minimum	-35.08	-24.63	-13.32	-10.42	-3.03	-8.30	0.06	-7.55	-15.70	-13.51	-29.18	-13.92	-25.84	-22.33
Std. Deviation	8.10	9.50	3.92	1.55	0.94	3.61	0.14	2.22	6.27	4.55	6.96	6.00	7.20	5.84
Sharpe Ratio	0.12	0.16	0.09	0.10	-0.31	0.00		0.22	0.08	0.16	0.05	0.09	0.10	0.18
Skewness	0.08	-0.11	-0.68	-2.04	-0.61	0.31	-0.56	-0.05	0.39	-0.11	-0.30	-0.01	-0.74	-0.85
Kurtosis	4.08	0.54	0.84	12.74	0.55	-0.02	-1.18	2.50	3.29	1.50	2.13	-0.13	1.91	2.73
JB	117.89	1.37	17.24	4.34	3.44	2.66	19.67	43.15	88.10	17.35	41.39	0.12	3.72	3.34
Prob.	0.00	0.50	0.00	0.18	0.18	0.27	0.00	0.00	0.00	0.00	0.00	0.94	0.16	0.19
Obs.	169	169	169	169	169	169	169	169	169	169	169	169	84	49
Yearly Statistic														
Mean	15.29	22.64	7.86	5.40	0.37	3.69	3.88	9.57	9.83	12.83	8.01	10.55	10.97	19.15
Std. Deviation	28.07	32.89	13.57	5.43	3.26	12.51	0.49	7.68	21.71	15.77	24.11	20.77	24.94	25.39
Sharpe Ratio	0.41	0.57	0.29	0.34	-1.07	-0.02		0.75	0.27	0.57	0.17	0.32	0.33	0.63
Downrisk Excess Mean	-12.91	-9.94	-16.54	-0.30	-1.52	-1.15		-3.32	-16.89	-7.08	-23.21	0.62	-23.85	-10.66
Downrisk β	1.19	1.17	1.00	0.07	-0.04	-0.01		0.32	1.14	0.53	1.33	0.22	1.55	1.90

Panel B - CAPM

Monthly Statistic	Renewable Energy	Cleantech	Stocks	Corporate Bonds	Government 30Y Bonds	Government 30Y Bonds	3 Month U.S. T Bill	Hedge Funds	Private Equity	Real Estate	Emerging Markets	Commo-dities	Wilderhill	S&P Clean
Intercept	0.0058	0.0119		0.0013	-0.0028	-0.0003		0.0037	0.0013	0.0058	-0.0008	0.0050	0.0068	0.086
	0.2639	*0.0629*		*0.2539*	*0.0001*	*0.9250*		*0.0093*	*0.7040*	*0.0637*	*0.8153*	*0.2736*	*0.0737*	*0.2313*
β	1.1220	1.1253		0.0521	-0.0266	0.0198		0.2882	1.0834	0.5156	1.2947	0.1834	1.5632	1.5864
	0.0000	*0.0000*		*0.0797*	*0.1445*	*0.7767*		*0.0000*	*0.000*	*0.0000*	*0.0000*	*0.1127*	*0.0000*	*0.0000*
Adjusted R²	0.2879	0.2103		0.0118	0.0064	-0.0052		0.2593	0.46	0.1917	0.5233	0.0086	0.7472	0.4636
F-statistic	72.5633	48.1626		3.1058	2.1488			62.9611	150.3893	42.9746	195.3460	2.5418	273.0138	50.2604
S.E. of regression	0.0686	0.0845		0.0154	0.0095	0.0363		0.0189	0.0460	0.0410	0.0483	0.0599	0.0363	0.0539

Source: The Renewable Energy Index result from own calculations. The World Alternative Energy Index (WAEX), the Morgan Stanley Capital Index World Index (Stocks), the J.P. Morgan Global Government Bond index (Bonds), the C.S/Tremont Hedge Fund Index (hedge funds), the Liquid Private equity Fund 50 (Private Equity), the World Real Estate Investment Trust Index (Real Estate), the Dow Jones Emerging Market Index (Emerging Markets), the Goldman Sachs Commodity Index (Commodities), the Wilderhill New Energy Global Innovation Index (Wilderhill) and the S&P Clean Energy Index (S&P Clean) are obtained from Thomson Financial Datastream. The Three Month U.S. T-Bill is obtained from St Louis FRED.

Table 4
Correlation Matrix

Table 4: exhibits the correlations between all asset classes based on simple returns. The number of observations is equal to all asset classes, that is the reason why the Wilderhill Index and the S&P Clean Energy Index are omitted. See table 2, Data Description for more information on each asset class.

	Renewable Energy	Cleantech	Stocks	Corporate Bonds	Government Bonds	Government 30Y Bonds	3 Month U.S. T Bill	Hedge Funds	Private Equity	Real Estate	Emerging Markets	Commo-dities
Renewable Energy Index	1.00	0.29	0.54	0.05	-0.17	-0.09	-0.05	0.31	0.42	0.27	0.69	0.11
Cleantech	0.29	1.00	0.47	0.01	-0.23	-0.11	0.09	0.33	0.42	0.20	0.43	0.26
Stocks	0.54	0.47	1.00	0.14	-0.10	0.02	0.09	0.52	0.68	0.45	0.72	0.12
Corporate Bonds	0.05	0.01	0.14	1.00	0.70	0.57	0.11	0.28	-0.03	0.21	0.06	0.14
Government Bonds	-0.17	-0.23	-0.10	0.70	1.00	0.68	0.02	0.07	-0.20	0.11	-0.16	-0.01
Government 30Y Bonds	-0.09	-0.11	0.02	0.57	0.68	1.00	-0.11	-0.01	-0.28	0.02	-0.09	0.11
3 Month U.S. T-Bill	-0.05	0.09	0.09	0.11	0.02	-0.11	1.00	0.16	0.12	0.07	-0.07	-0.04
Hedge Funds	0.31	0.33	0.52	0.28	0.07	-0.01	0.16	1.00	0.50	0.30	0.47	0.25
Private Equity	0.42	0.42	0.68	-0.03	-0.20	-0.28	0.12	0.50	1.00	0.30	0.61	0.10
Real Estate	0.27	0.20	0.45	0.21	0.11	0.02	0.07	0.27	0.30	1.00	0.38	-0.03
Emerging Markets	0.69	0.43	0.72	0.06	-0.16	-0.09	-0.07	0.47	0.61	0.38	1.00	0.19
Commodities	0.11	0.26	0.12	0.14	-0.01	0.11	-0.04	0.25	0.10	-0.03	0.19	1.00

Source: The Renewable Energy Index result from own calculations. The World Alternative Energy Index (WAEX), the Morgan Stanley Capital Index World Index (Stocks), the J.P. Morgan Global Government Bond index (Bonds), the C.S/Tremont Hedge Fund Index (hedge funds), the Liquid Private equity Fund 50 (Private Equity), the World Real Estate Investment Trust Index (Real Estate), the Dow Jones Emerging Market Index (Emerging Markets) and the Goldman Sachs Commodity Index (Commodities) are obtained from Thomson Financial Datastream. The Three Month U.S. T-Bill is obtained from St Louis FRED.

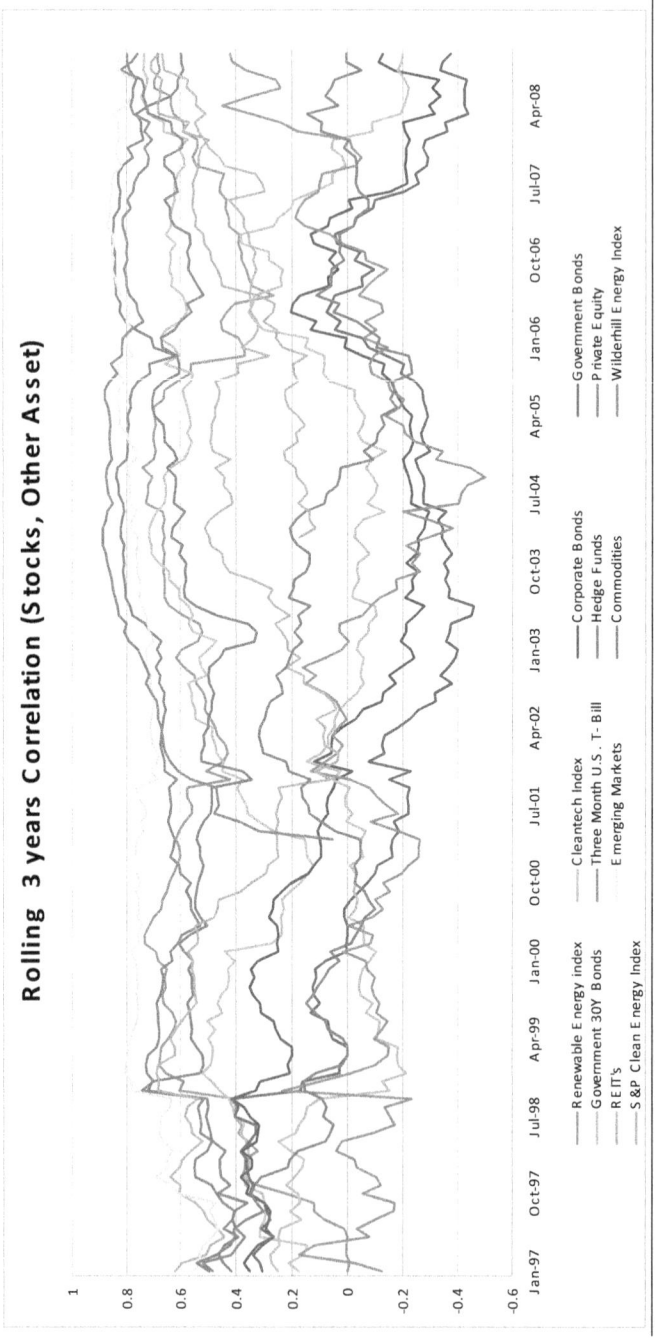

Figure 3a: Three Years Rolling Correlation Between Stocks and the Other Assets

The figure exhibits the rolling three year correlations between stocks represented by the MSCI World Index and all other asset classes used in this analysis based on simple returns. The number of observations is equal to all asset classes. See Data Description for more information on each asset class. *Source:* The Renewable Energy Index result from own calculations. The World Alternative Energy Index (WAEX), the Morgan Stanley Capital Index World Index (Stocks), the J.P. Morgan Global Government Bond index (Bonds), the C.S/Tremont Hedge Fund Index (hedge funds), the Liquid Private equity Fund 50 (Private Equity), the World Real Estate Investment Trust Index (Real Estate), the Dow Jones Emerging Market Index (Emerging Markets), the Goldman Sachs Commodity Index (Commodities), the Wilderhill New Energy Global Innovation Index (Wilderhill) and the S&P Clean Energy Index (S&P Clean) are obtained from Thomson Financial Datastream. The Three Month U.S. T-Bill is obtained from St Louis FRED.

There are certain interesting changes through time. For example, the correlation of stocks and the Renewable Energy Index gets higher as more companies enter the index, the index behaves more and more like the stock market. The correlation has long been fairly constant around 0.6 till three years ago, when the correlation first fell sharply to 0.3 and rising to its current level of 0.7. This is again the boom initiated in 2003 where it outperformed the general stock market. The last two years the Renewable Energy Index stabilised and crashed, just as the stock market, which explains the rising correlation.

Overall one can observe that correlations lie at a higher level than in 1997. Longin and Solnik (2001) and Ang and Chen (2002) concluded that during bear markets correlations of international equities are much higher than during tranquil or bull markets. And the past eight years have indeed shown several bear market periods and Figure 3 shows that correlation increases during these periods. Another conclusion drawn was that correlation asymmetries are higher for extreme downward movements of stock returns. When looking at Figure 3 greater asymmetries can be observed when focussing at the past six months. When stock returns plumbed all over the globe, divergence widen between alternatives (except commodities) and bonds.

The correlations of the alternative asset classes seem to converge to the same level during the last three years. Conversely, the correlation of stocks and bonds diverges from this trend to a higher negative level. This divergence was also noticed by Wong and Vlaar (2003). They also detected a *Flight to quality*, during strong stock market turmoil investors try to escape the sharp fall in prices and buy into bonds this enlarges the negative correlation. This Flight to quality is clearly seen in Figure 3, where bonds increasingly operate as substitutes for stocks.

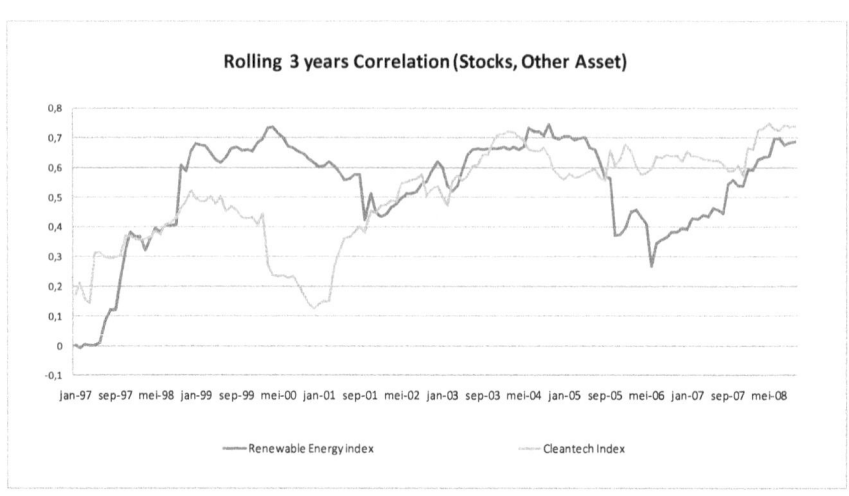

Figure 3b: Three Years Rolling Correlation Between Stocks and Renewable Energy or Cleantech
The figure exhibits the rolling three year correlations between stocks represented by the MSCI World Index and renewable energy represented by the Renewable Energy Index as well as stocks with cleantech represented by the WAEX index, based on simple returns. The number of observations is equal to all asset classes. See Data Description for more information on each asset class. *Source*: The Renewable Energy Index result from own calculations. The World Alternative Energy Index (WAEX) and the Morgan Stanley Capital Index World Index (Stocks) are obtained from Thomson Financial Datastream.

C.   Portfolio Optimization

The optimization of the portfolios is done under a mean/variance framework as explained in the methodology section. The optimization was performed without posing any restrictions. All asset classes are tradable directly or via replicating funds or ETF's (Exchange Traded Funds) so short selling is possible and therefore such a restriction is not needed. The portfolios are constructed by weighting each asset class randomly 5,000 times, plotting the return and standard deviation of each random drawn portfolio. This way creating a cloud of feasible portfolios. The efficient frontier lies on the right upper edge of the cloud of feasible portfolios. All portfolios in this study include the assets: stocks, corporate bonds, government bonds, 30 year government bonds, hedge funds, private equity, real estate, emerging markets and commodities. The inclusion of renewable energy and/or cleantech is mentioned explicitly for each set of portfolios. Figure 4 exhibits the feasible portfolios of two different sets and the average mean and standard deviation of stocks as a point of reference. The green set shows the feasible portfolios including the assets Renewable Energy and cleantech. The red set displays the feasible portfolios excluding renewable energy and cleantech. The efficient

36

frontiers of both sets are displayed in the corresponding colours. The reference point stocks is shown in light blue.

Figure 4 clearly shows the red cloud in front of the green cloud. The efficient frontiers intersect at an annually mean return of 8.8 percent given a standard deviation of 4.4 percent. Above the 8.8 percent return level the green line exceeds the red line, meaning the frontier of portfolios including renewable energy and cleantech is more efficient than the frontier of portfolios excluding both sectors. Below the 8.8 percent return level it is the other way around. Meaning, for a given return below the 8.8 percent, a lower standard deviation can be achieved when investing in a portfolio without renewable energy and cleantech.

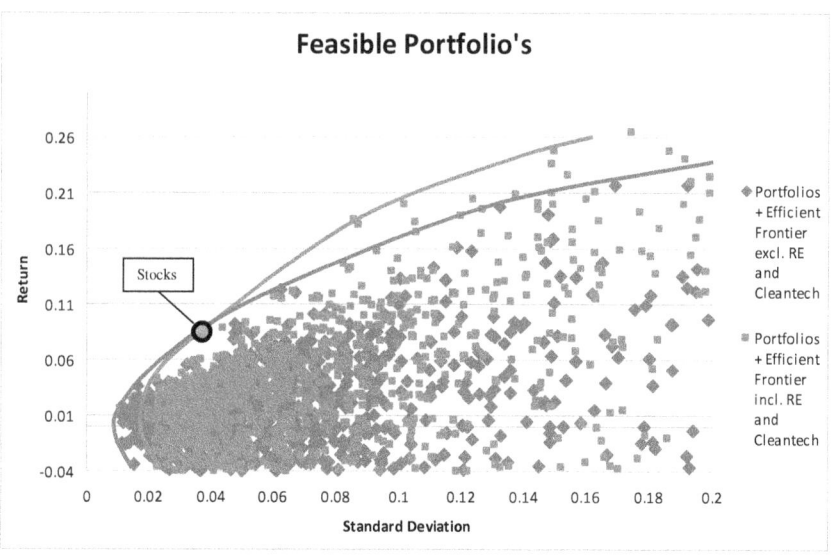

Figure 4: Portfolio Optimization With Two Kinds of Portfolios, One With Renewable Energy and Cleantech Together With the Other Asset Classes and One Without Renewable Energy and Cleantech.
Figure 4: displays all feasible portfolios of two different sets. As a reference point the average return and standard deviation of stocks are shown in light blue. The green set shows all feasible portfolios including the assets Renewable Energy and Cleantech. The red set displays all feasible portfolios excluding Renewable Energy and Cleantech. The efficient frontier lies on the edge of the set of feasible portfolios. The efficient frontier including Renewable Energy and Cleantech is displayed in green. The efficient frontier excluding Renewable Energy and Cleantech is shown in red. The portfolios are constructed by weighting each asset class randomly 5,000 times, plotting the return and standard deviation of the portfolio each random draw. Creating a set of feasible portfolios. The efficient frontiers intersect at an annually mean return of 8.8 percent given a standard deviation of 4.4 percent. Thus, riskavers investors preferring a lower return than 8.8 percent should invest in a portfolio excluding the assets Renewable Energy and Cleantech. Investors preferring a return over 8.8 percent should invest in a portfolio including the assets Renewable Energy and Cleantech.

As a result, risk avers investors aiming at a lower average annual return than 8.8 percent should invest in a portfolio excluding the assets renewable energy and cleantech. Investors that are more risk tolerant and preferring an average return over 8.8 percent should invest in a portfolio including the assets renewable energy and cleantech. The part in figure 4 where the green line exceeds the red line, illustrates the added value of including renewable energy and cleantech to the portfolio. A portfolio with renewable energy and cleantech outperforming a traditional portfolio seems plausible, given their high Sharpe Ratio and good return/risk performance illustrated in Figure 2.

Figure 5 exhibits a set of feasible portfolios including the asset class renewable energy displayed in dark green and a set of portfolios including cleantech displayed in light green and stocks as a reference point. The dark green efficient frontier exceeds the light green frontier. Portfolios including renewable energy outperform portfolios including cleantech. Implying that for a given standard deviation a higher return can be achieved by adding renewable energy to your investment selection instead of cleantech.

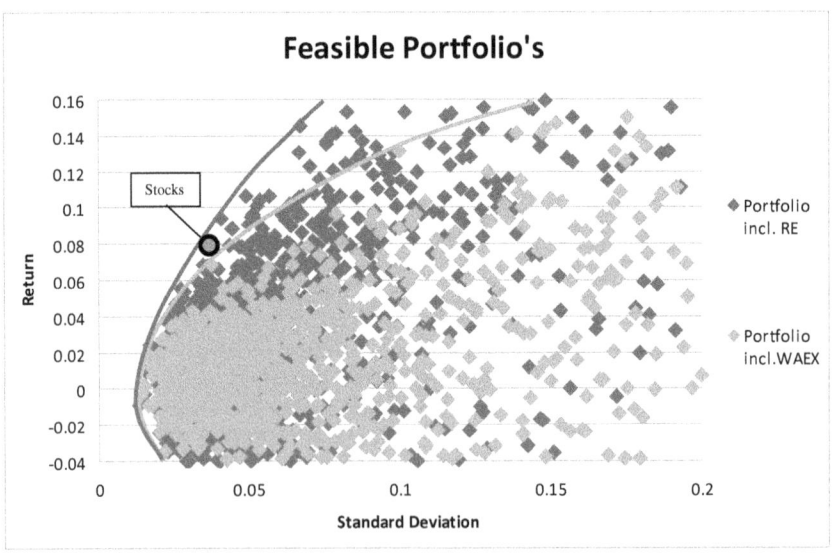

Figure 5: Portfolio Optimization With Two Kinds of Portfolios, One With Renewable Energy and the Other Asset Classes and One With Cleantech and the Other Asset Classes.
Figure 5: displays all feasible portfolios of two different sets. As a reference point the average return and standard deviation of stocks are shown in light blue. The dark green set shows all feasible portfolios only including the asset Renewable Energy and not the asset Cleantech. The light green set displays all feasible portfolios including only Cleantech and not Renewable Energy. The efficient frontier of the portfolios including Renewable Energy is displayed in dark green and the portfolios including Cleantech is displayed in light green. The portfolios are constructed by weighting each asset class randomly 5,000 times, plotting the return and standard deviation of the portfolio each random draw. Creating a set of feasible portfolios. The efficient frontiers do not intersect. The portfolios including the asset Renewable Energy outperform the portfolios including the asset Cleantech, with a greater return given a certain standard deviation.

This is somewhat unexpected, regarding the higher Sharpe Ratio of cleantech. The covariances of renewable energy provide an advantage over cleantech in a portfolio. The interrelations of the assets make the overall portfolio achieve a higher return, which is easily observed in Figure 5. As Cambell (2004) pointed out, a low or negative correlation with other asset classes makes an asset a beneficial investment vehicle for an investors' portfolio. However, the correlations of renewable energy are not very low, apparently low enough for a portfolios performance to improve.

Which portfolio is suitable for a particular investor depends on where the investors personal utility curve touches the efficient frontier. This tangency point determines the optimal allocation for that investor. Each investor knows its own utility and can reason if renewable energy and cleantech should be included or not. However, bear in mind the relatively short data range and that outcomes are based on historical results, so caution is needed when interpreting these results.

After performing the risk/return analysis, discussing the correlations, the correlations over time, a portfolio optimization. The empirical findings are discussed comprehensively and the research question why to invest in renewable energy or cleantech can be answered in the following concluding section.

5.   Conclusion

The final section of this thesis consists of the conclusion, followed by suggestions for further research and finally a discussion about the future of renewable energy and cleantech is given in the outlook.

A.   Concluding Remarks

In this thesis the renewable energy and the cleantech sector are analysed extensively for the period 1994 to 2008 (ytd) with monthly data resulting in 169 observations. This study is relevant for portfolio managers, institutional investors, but also to investors in general or people interested in renewable energy. Purpose of this thesis is to inform investors if investing in renewable energy is really different than investing in 'traditional' assets, as often claimed in the media. Both sectors make a valuable contribution to an existing portfolio of 'traditional' assets, when investors do not mind a portfolio with a higher standard deviation of

4.4 percent. When they do? They are better off with a portfolio excluding renewable energy and cleantech, holding only traditional assets. Thus risk neutral investors wanting a return exceeding 8.8% on their portfolio should include renewable energy and cleantech. When studied apart, within a portfolio perspective, renewable energy clearly outperforms cleantech as a potential alternative. The generated efficient frontiers clearly show these outcomes in figure 4 and 5. The added value of renewable energy and cleantech is evident. When the risk of a portfolio exceeds the 4.4 percent, a higher return can be achieved with inclusion of both sectors in the portfolio. These results seem to point out that renewable energy and cleantech have a high risk profile, riskier than the average stock. Their profile more matches those of emerging market- and technology stocks. When investing in renewable energy / cleantech stocks one invests in growth stocks with high potential. In time these stocks will convert more to general stocks. Their volatility will decrease and consequently their correlation with stocks will increase. However, for now it is a time to include these stocks in your portfolio before prices will go up again and returns will level out with ordinary stocks. This does not mean that the long term added value of renewable energy and cleantech is the same as the average stock in the market. On the contrary, the political support, the growing green awareness in the public, the climate problems, the dislike for oil-dependence, all together form fundamentals in which the market for renewable energy and cleantech stocks will flourish, even when these markets mature.

During the last 15 years both sectors shown strong growth. However, caution is needed with interpreting the following results, because of the relative short data range covering multiple drastic economic events. Renewable energy earned an annualized average return of 15 percent per year for the last fifteen years with a standard deviation of 28 percent. Resulting in a Sharpe Ratio of 0.41. Relative to stocks, with a return of 8 percent (14 percent standard deviation), renewable energy performed really well. Renewable energy has a correlation of 0.54 with stocks, a market beta of 1.12 (adjusted R^2 of 29 percent) and a downside risk beta of 1.19, with the MSCI World Index representing the global stock market. Indicating downward movements of the market are a followed by slightly stronger movements down of renewable energy and movements up are followed less strongly. Cleantechs returns are somewhat different, a return of 23 percent with 33 percent standard deviation, resulting in a higher Sharpe Ratio of 0.57. Cleantechs correlation with stocks is 0.47. With a market beta is of 1.12 (adjusted R^2 of 21 percent) and a downside risk beta of 1.17 cleantech shows similar market relations as renewable energy. Not surprisingly, since one is a subsector of the other.

Regarding the downside risk betas renewable energy and cleantech are no good alternatives in case of a downturn in the stock market. However, correlations with other asset classes are different for each sector. Renewable energy has no correlations higher than 42 percent, except with emerging markets (0.69) and cleantech has no correlations higher than 43 percent with the other asset classes. This gives them a fairly distinctive characterization, raising the notion that they are a valuable addition to existing investment vehicles. Not due to their diversification characteristics but because they are growth stocks with high potential. This is confirmed by improving the efficient frontier with inclusion, whereby renewable energy surpasses cleantech as a potential asset class.

B.   Further Research

This thesis focuses on listed indices only, assuming the displayed indices make a useful representation of the discussed asset classes. Especially for renewable energy, hedge funds and private equity, investors are also (or even more) interested in the characterization of the non-listed part of these asset classes and the implications these characteristics induce for their portfolios. If one can overcome the data shortage of non-listed fund or project performance, or one has access to those performances in some form or other, this can be of great value and will be appreciated throughout the investment sector. And therefore, it remains a worthy topic for further research.

Over time, more companies will produce more renewable energy and an increasing number of companies will adhere to the inclusion criteria. This study can than be reviewed when the Renewable Energy Index holds more constituents, increasing the representative power of the Renewable Energy Index. The second problem will also be mitigated because of availability of a longer data period. In this way, economic shocks have less impact on the overall results of the performance and more weight could be given to the outcome.

The companies analysed for inclusion in the Renewable Energy Index are all companies that (also) report in English. Companies that report in other languages are not analysed, this is a short coming in this study. Although presumable not many companies are overlooked this way. Asia, Afrika and Latin America may hold companies that adhere to the inclusion criteria. Especially China may hold pure play renewable energy companies. Analysing these markets can be quite relevant for further research.

Another suggestion is to time weight the portfolio optimization. Giving more power to recent observations. Two time weight return schemes have been suggested in the literature. One is a form of the Fisher distributed lag model implemented by Ray and Nawrocki (1996) and the second is the "half-life" weighting scheme suggested by Sharpe (1995). Both schemes can be justified. The first reason is simplicity, they are easy to implement because the weighting is based on the number of time periods used. Secondly, it seems reasonable that investors adapt their views to recent observations and that the more distant the experience, the lesser effect such information has in forming those expectations (Lee and Stevenson, 2003).

Historical data can also be weighted by better forecasting the future volatility. This can be done with exponentially weighted moving average (EWMA) techniques or generalized autoregressive conditional heteroscedastic (GARCH) models. Lee and Stevenson (2003) discuss these methods and they choose to 'simply' time weight the returns data.

C.   Outlook

The future of renewable energy looks extremely promising. The question is how long it will take before the majority of our power needs are generated from renewable sources. According to the EIA this will happen in 2050. This depends on certain factors. Overall energy demand rises due to global population growth and increasing wealth. Emerging markets will ensure a significant increase in the demand for energy. The demand for renewable energy is triggered by a growing awareness that a green environment is essential for a sustainable future, signals of climate change and too much emissions contribute to that awareness. This awareness sets renewable energy high on the political agenda. The need to lower emissions and to lower dependence on fossil fuels let policy makers incorporate renewable energy in their mission statements, which result in targets, like the 20% renewable energy production in 2020 target set by the European Union. This plan wants to stimulate member states to reach an average of 20% renewable energy of the total energy use by the year 2020. Other countries have similar targets or are structuring such plans at the moment.

When constructing those plans certain facts should be taken into account. Governments should take collective actions in passing laws to ensure renewable energy has priority before grey energy when going to the grid. Government should support RE in its infancy so it can grow from beneath the shadow of the fossil fuel empire. This is done for instance by feed in laws, Renewable Energy Certificate System or other local variants. Feed in laws are regarded as the most successful, with the UK and Germany setting good examples. The feed in laws should gradually be lowered when RE is strong enough to stand alone. What a government should never do is what Dutch policy makers did in the past few years. They changed their policy regularly in a few years time. In stead of creating a stable environment to give seed capital a chance, there policy was more like a traffic light, what was killing for investments, by setting subsidies to a stop and changing the rules almost each year. A stable environment can never be achieved acting like this.

However, more promising is the technological progress the sector has made. In this way it can stand alone without subsidies in a few years. Wind is already profitable, almost without feed in tariffs in certain favourable places, just like solar PV and CSP (Concentrated Solar Power). For wind cost effective areas are coastal areas, on- and offshore, particularly those with constant wind. For solar these areas are in California, U.S. and the Mediterranean. It will not take long before solar will hit grid parity, meaning the cost of solar generated power is equal to the cost of traditional fossil based generated power. On this basis, it is clear that we are not there yet. Even with accelerated rates of investment in energy R&D, we should be focusing on mitigating greenhouse gas emissions, and reducing energy use through simple conservation coupled to the modification of existing energy technologies to perform cleaner and more efficiently during the current 20 to 50 year lag period, prior to full scale transitioning to new energy technologies. Perhaps ironically, developing countries may actually be in a position to transition to more sustainable energy technologies faster than in developed countries, because they lack a complex, entangled energy infrastructure. In any case, such a transition may be only feasible with substantial assistance and technical support of the developed world.

The reason why renewable energy technologies do not comprise the majority of installed capacity is because of easy accessible and therefore cheap fossil alternatives, but more importantly the "slow capital stock turnover", or SCST. SCST relates to the typically long life span, 50 years for power plants, of older technologies. Thus, unless there is a regulatory phase out of these old systems, such as the recent U.S. Clean Air Act to replace polluting coal fired power plants, there will be a significant lag period between implementation of renewable energy technologies and their full scale commercialization and adoption in every day life (Ogden, 1999 and Painuly, 2001).

My personal view on renewable energy 50%+ is somewhat more optimistic, seeing the speed of technological progress. The sector is highly profitable and will stay this way due to 20% RE targets for the coming years. But what after 2020? Solar power will be sufficient to power every home, vehicles won't need fossil fuels, maybe flying will take a bit longer to replace fossil fuelled combustion and companies with their energy intense production will be dependant on fossil based power longer than individual households. The point is demand for power from original power companies will diminish in the long run and the question is: how will those companies react to this shift from monopolistic competition to perfect competition or no business at all? Because a shift is now already visible that the next generation energy source will not be developed by the present utility companies but by electronic companies investing heavily in solar powered solutions and artificial photosynthesis-like alternatives. This will cause a shift in providing energy from energy companies to technology companies or even consumers themselves, who will be able to generate renewable power on an individual level. In general renewable energy will be, for a big part, generated more decentralized. For example through efficient thin film painted anywhere, or small cogeneration facilities, also known as combined heat and power (CHP), installed in homes, where the use of a heat engine or a power station simultaneously generates both electricity and useful heat. In any way, cleantech will be able to create more suitable solutions worldwide, acceptable and affordable for households, so the future looks quite promising.

Appendix I

The log prices of the asset classes.

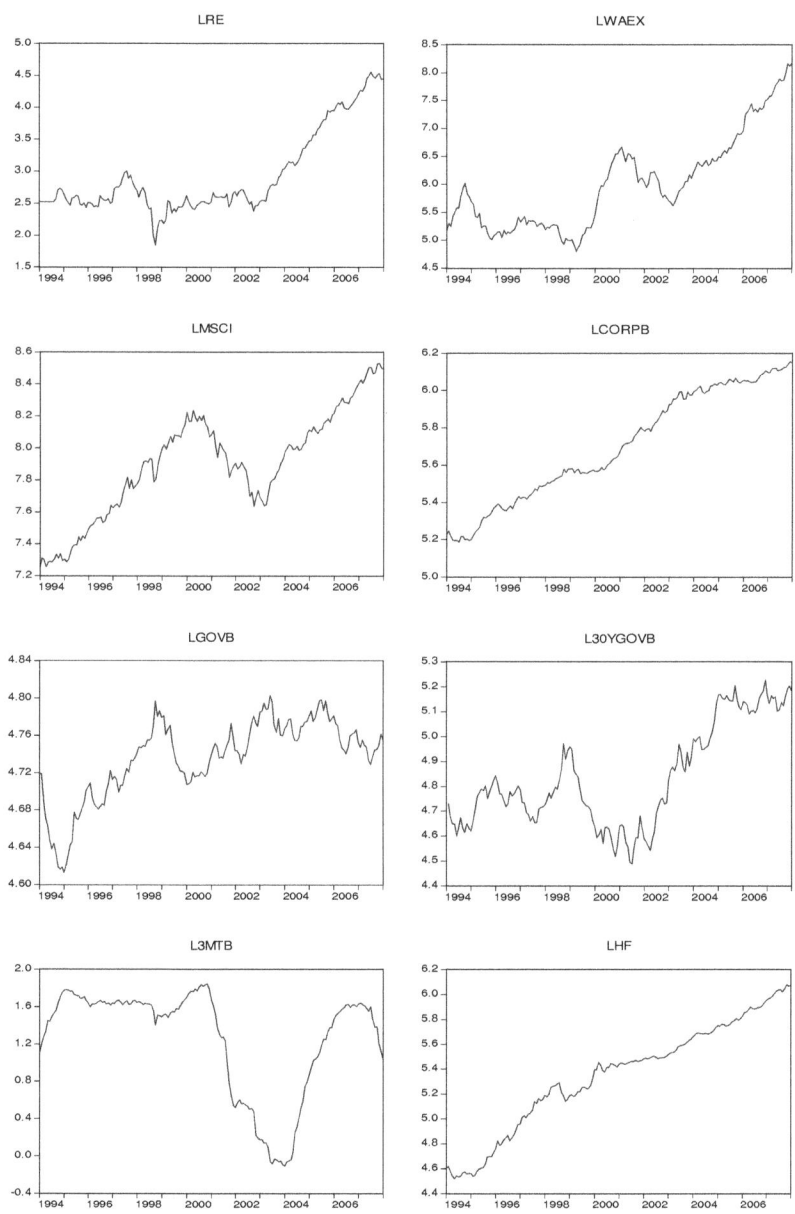

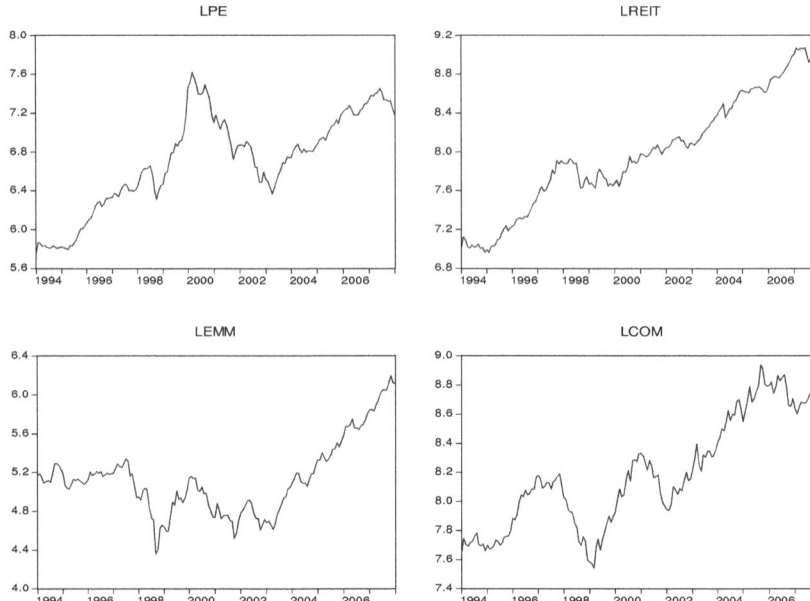

Appendix II

Correlation between the 10 MSCI Subsectors and the Renewable energy index and the Cleantech Index

	Energy	Materials	Ind.	Cons. D.	Cons. S.	Health C.	Financials	IT	Telecom	Utilities	Ren. E.	Cleantech
Energy	1.00											
Materials	0.67	1.00										
Industrials	0.58	0.82	1.00									
Cons. Discretionary	0.46	0.71	0.88	1.00								
Cons. Staples	0.37	0.43	0.54	0.48	1.00							
Health Care	0.32	0.29	0.46	0.42	0.65	1.00						
Financials	0.42	0.63	0.80	0.79	0.65	0.55	1.00					
IT	0.37	0.52	0.72	0.80	0.24	0.33	0.56	1.00				
Telecom	0.25	0.39	0.57	0.71	0.30	0.37	0.54	0.72	1.00			
Utilities	0.56	0.45	0.54	0.46	0.59	0.48	0.54	0.27	0.35	1.00		
Renewable Energy	0.40	0.46	0.54	0.49	0.39	0.29	0.56	0.40	0.30	0.29	1.00	
Cleantech	0.35	0.45	0.52	0.44	0.26	0.20	0.36	0.43	0.31	0.41	0.30	1.00

Correlation between the sub-subsectors of the MSCI Industrials and the Renewable energy index and the Cleantech Index

	Cap. G.	Com. S.	Transp.	Ren. E.	Cleantech
Capital Goods	1.00				
Commercial Services	0.80	1.00			
Transport	0.80	0.76	1.00		
Renewable Energy	0.53	0.49	0.49	1.00	
Cleantech	0.53	0.41	0.40	0.30	1.00

Appendix III

Pure players	Ticker	Type	Country	Exchange	Description
Novera Energy PLC	NVE	H,W,B	UK	LON	Generates renewable power at 46 landfill gas, 10 hydro power stations
Actelios SpA	ACT	B,S	IT	ISE	Renewable energy producer, mostly biomass based
Babcock & Brown Wind Partners	BBW	W	AUS	WBAG	Wind developer
Iberdrola Renovables	IBR	H,W	ES	MSE	Renewable energy 96% wind and 4% small hydro
Cia Paranaense de Energias	ADR	H	BRA	NYSE	17 hydroelectric (18,000 MW) and one thermoelectric power plants (69MW)
Empresa Nacional de Energias SA	ADR	H	Chile	NYSE	Endesa Chile mostly Hydro power 12,720 MW
Environmental Power Corp.	EPG	B	US	NASDAQ	RE producer, produces NG, generates electricity from biomass
Polaris Geothermal Inc	GEO	G	CAN	TSX	Geothermal energy producer
Boralex Inc	BLX	H,W,B	CAN	TSE	Private electricity producer mostly from biomass, combined combustion
Canadian Hydro Development	KHD	H,W,B	CAN	TSX	Developer and constructor and operator
Greentech Energy System	GES	W	DEN	OMX	Developer and constructor and operator
AES Tiete SA	GETI4	H	BRA	SAO	Production of electric energy, 10 hydroelectric power plants, 2,651 MW
Cia Energetica de Sao	CESP6	H	BRA	SAO	Production electric energy, 6 hydro plants 7,456 MW
Contact Energy Ltd	CEN	H,W,G	NZ	NZE	Generator of hydro and geothermal electricity
Western Wind Energy	WND	W	CAN	TSX	Wind developer, with 3700 acres 100% owned land

Pure players	Sales Eur mln	RE Sales	% RE cap	Prod. MW end 2007	RE prod. GWh end 2007	% RE prod	H%	W%	B%	G%	M%
Novera Energy PLC	46.8	43.5	93.0%		564.0	100.0%	8.7%	5.0%	86.3%	0.0%	0.0%
Actelios SpA	90.6	63.2	69.7%	1,500.0	12,000.0	100.0%	0.0%	0.0%	100.0%	0.0%	0.0%
Babcock & Brown Wind Partners	103.7	55.8	53.8%	1,860.0	5,376.9	100.0%	0.0%	100.0%	0.0%	0.0%	0.0%
Iberdrola Renovables	1,580.0	1,040.3	65.8%	7,704.0	14,708.0	100.0%	3.7%	96.3%	0.0%	0.0%	0.0%
Cia Paranaense de Energias	5,387.4	4,115.0	76.4%	5,152.0	18,064.6	99.6%	100.0%	0.0%	0.0%	0.0%	0.0%
Empresa Nacional de Energias SA	2,364.5	1,484.9	62.8%	12,720.0	31,705.2	62.8%	99.7%	0.3%	0.0%	0.0%	0.0%
Environmental Power Corp.	35.1		0.0%		204.9	100.0%	0.0%	0.0%	100.0%	0.0%	0.0%
Polaris Geothermal Inc	3.1	2.7	86.7%	10.0	85.0	100.0%	0.0%	0.0%	0.0%	100.0%	0.0%
Boralex Inc	113.1	102.9	91.0%	351.0	2,082.1	96.1%	10.2%	11.4%	78.4%	0.0%	0.0%
Canadian Hydro Development	43.8	43.8	100.0%	363.8	921.7	100.0%	39.9%	46.7%	13.4%	0.0%	0.0%
Greentech Energy System	6.9	6.9	100.0%	68.3	80.6	100.0%	0.0%	100.0%	0.0%	0.0%	0.0%
AES Tiete SA	586.7	586.7	100.0%	2,651.0	13,185.0	100.0%	100.0%	0.0%	0.0%	0.0%	0.0%
Cia Energetica de Sao	1,006.8	1,006.8	100.0%	7,456.0	32,845.0	100.0%	0.0%	100.0%	0.0%	0.0%	0.0%
Contact Energy Ltd	963.4	447.1	46.4%		5,625.0	51.0%	64.7%	0.0%	0.0%	35.3%	0.0%
Western Wind Energy	1.1	1.1	100.0%	35.0	77.0	100.0%	0.0%	100.0%	0.0%	0.0%	0.0%

Operating time (average hours)	h / year	Cap. Factor	Exchange rates on 31-12-07	
Hydro (H)	8500	0.97	UK pound - Euro	0.74
Wind onshore (W)	2200	0.25		
Wind offshore (W)	3000	0.34	USD - Euro	1.47
Biomass (B)	8000	0.91		
Geothermal (G)	8500	0.97	CAN - Euro	1.44
Tidal / Wave (M)	5000	0.57		
Solar (S)	1200	0.14	DK – Euro	7.46
			Real - Euro	2.61
			NZ - Euro	2.07

Appendix IV

HF 2.D. Johansen Co-integration and Granger Causality

As already raised in the introduction, there are somewhat more advanced measures, aside
from correlation measures, to indicate a relation between two series. For example, Co-
integration (Johansen, 1991) and Causality (Granger, 1987). Co-integration is a measure
which tests whether, in general terms, two series co-move. The difference with correlation is
that when two series are correlated, the series move up or down synchronically per period. In
contrast to co-integrated series, which cannot drift in opposite directions for multiple periods
without coming back to a mean distance eventually. Although it does not mean that on a
monthly basis the two series have to move in synchrony at all time. Testing for co-integration
the relation of renewable energy with other asset classes can be tested further. The series
could on the long run move in similar ways, because they might be subjective to the same
influences and will thus both be bound to certain co-movements. If prices are mean reverting,

asset prices are bound together in the long term, by a common stochastic trend. When this is the case, prices are co-integrated (Alexander, 2001). There have been several methods of computing co-integration. No elaboration will take place on these other methods. The method to be applied here is that of Johansen (1991)[7] in line with Nobel laureates Engle and Granger (1987)[8], which established a two-step procedure, which is necessary when testing for co-integration. Most series follow a unit root process integrated of order 1:

$$y_t \sim I(1)$$

also called a non-stationary process or a random walk. The use of non-stationary series can lead to spurious regressions and bad modelling in general. It is possible that two series completely unrelated could statistically appear to be related. The test is performed through the Augmented Dickey Fuller Test (Engle and Granger, 1987) and states the following hypotheses:

$$H_0: \emptyset = 1 \ in \ y_t = \emptyset y_{t-1} + u_t$$
$$H_1: \emptyset < 1 \ in \ y_t = \emptyset y_{t-1} + u_t$$

When unit root is not rejected one must take the first difference $(\Delta y_t = y_t - y_{t-1})$ of the series and perform the test again. As mentioned, most series are only integrated of order one and will therefore no longer follow a unit root process. However, when it does the series is integrated of order two (or potentially higher) and will therefore be omitted from the co-integration analysis. In this case the appeal of a *VECM* formulation is that it combines flexibility in dynamic specification with desirable long-run properties. The two-equation V.A.R. is specified as a Vector Error Correction Model (VECM) in the following way:

$$y_t = \alpha_1 + \delta_1 \left(\sum_{i=1}^{n} \gamma_i y_{t-i} \right) + \sum_{i=1}^{n} \beta_i x_{t-i} + \varepsilon_{1,t}$$

---

[7] This method is especially appealing since it provides a unified framework for estimating and testing co-integrating relations in the context of a VECM model. Thus, all the variables are treated as endogenous, this approach avoids the arbitrary choice of the dependent variable in the co-integrating equations, as in the Engle-Granger methodology. They have also shown to have good large- and finite-sample properties (Phillips, 1991, Cheung and Lai, 1993).

[8] The Nobel Prize winners in Economics in 2003, Robert F. Engle "for methods of analyzing economic time series with time-varying volatility (ARCH)" together with Clive W.J. Granger "for methods of analyzing economic time series with common trends (co-integration)"

$$x_t = \alpha_2 + \delta_2 \left(\sum_{i=1}^{n} \gamma_i x_{t-i} \right) + \sum_{i=1}^{n} \beta_i y_{t-i} + \varepsilon_{2,t}$$

When specified as first difference:

$$\Delta y_t = \alpha_1 + \delta_1 \left(\sum_{i=1}^{n} \gamma_i \Delta y_{t-i} \right) + \sum_{i=1}^{n} \beta_i \Delta x_{t-i} + \varepsilon_{1,t}$$

$$\Delta x_t = \alpha_2 + \delta_2 \left(\sum_{i=1}^{n} \gamma_i \Delta x_{t-i} \right) + \sum_{i=1}^{n} \beta_i \Delta y_{t-i} + \varepsilon_{2,t}$$

where

Δy_t	equals the dependent variable in first differenced series
α	equals the intercept of the model
δ	equals the error correction coefficient
γ	equals the coefficient value for the lags within the error correction part of the V.A.R. model
β	equals the coefficient value for the other lags specified
ε	equals the disturbance term
n	equals for the number of lags
t	equals the time period

To test for significance both the Maximum eigenvalue and the Trace test of co-integration are determined. Both tests are defined in respective order as follows:

$$\lambda_{max}(r, r+1) = -T ln(1 - \hat{\lambda}_{r+1})$$

$$\lambda_{trace}(r) = -T \sum_{i=r+1}^{n} ln(1 - \hat{\lambda}_i)$$

where

r	equals the number of co-integrating vectors
$\hat{\lambda}$	equals the eigenvalue
T	equals the total number of time periods

These tests will be performed with E-views. The critical values for these tests will be presented in the empirical analysis.

As mentioned before, Granger causality measures whether one thing happens before another thing and helps predict it (Sorensen, 2005). The simple causal model is defined as follows:

$$x_t = \alpha_1 + \sum_{i=1}^{n} \beta_{1,i} x_{t-i} + \sum_{i=1}^{n} \gamma_{1,i} y_{t-i} + \varepsilon_{1,t}$$

$$y_t = \alpha_2 + \sum_{i=1}^{n} \beta_{2,i} x_{t-i} + \sum_{i=1}^{n} \gamma_{2,i} y_{t-i} + \varepsilon_{2,t}$$

Corresponding to the co-integration tests the model will be specified in first difference:

$$\Delta x_t = \alpha_1 + \sum_{i=1}^{n} \beta_{1,i} \Delta x_{t-i} + \sum_{i=1}^{n} \gamma_{1,i} \Delta y_{t-i} + \varepsilon_{1,t}$$

$$\Delta y_t = \alpha_2 + \sum_{i=1}^{n} \beta_{2,i} \Delta x_{t-i} + \sum_{i=1}^{n} \gamma_{2,i} \Delta y_{t-i} + \varepsilon_{2,t}$$

Definitions remain in line with the co-integration V.A.R.; only the correction mechanism is omitted. To test for significance between the causal relations of the variables the collective significance of the lagged variables will be tested through a model F-test. As presented in the next section.

HF 4.D. Unit Root Test

As mentioned earlier, there are enhanced measures to analyse co-movements between series. The following measures were used in this analysis. A Johansen co-integration test and a Granger causality test to see if there are long and short term relations, respectively. According to Engle and Granger (1987) the co-integration test consists of a two-step process. First, the unit root test is performed on the series and when a unit root process is found the test is performed on the differenced series, followed by step two, the co-integration test. When a series follows a unit root process or random walk, the series is omitted from further analysis; it is illogical in that case to test for co-integration. When analysing the graphs of all the series for stationarity (Appendix I), a clear trend is visible in almost all series. The three month U.S.

T-bill shows non-stationarity, as well as the government bonds for both maturities. It is not expected to find a unit root process in the rest of the series. Table 5 exhibits the results of the unit root test.

The unit root test indicates all series are integrated of order 1 (I (1)). Next the differenced series are tested for stationarity. Only the Three Month U.S T-Bill and Private Equity have a unit root process of the second order. All the other series reject both null hypothesis of having a unit root. The Three Month U.S T-Bill and Private Equity series will be omitted from the analysis due to a higher order of integration. All other series will be used for the co-integration and causality analysis

Table 5
Unit Root Tests

Table 5: exhibits the results of the Augmented Dickey Fuller unit root (random walk) tests for the different variables. For the optimal result, the Schwarz information criterion was used. The null and alternative hypotheses are H_0: unit root and H_1: no unit root respectively. The tests include an intercept and are computed through the ADF unit root test. The Three Month U.S. T-Bill and Private Equity will be excluded for further analysis. The variables are subdivided into several categories. All ranges are from 1994 to 2008 (ytd.). See table 2, Data description for more information on each asset class.

Asset	Probability[a]	
	Level	*First difference*
Renewable Energy Index	* 0.9891	0.0308
Cleantech	* 0.9725	0.0106
Market		
Stocks	* 0.7704	0.0408
Corporate Bonds	* 0.8797	0.0045
Government Bonds	* 0.5791	0.0008
Government 30Y Bonds	* 0.8728	0.0206
Three Month U.S. T-Bill	* 0.8491	** 0.2037
Alternatives		
Hedge Funds	* 0.8929	0.0092
Private Equity	* 0.4437	** 0.0503
Real Estate	* 0.8473	0.0214
Emerging Markets	* 0.9297	0.0427
Commodities	*0.8752	0.0094

a Probabilities are MacKinnon (1996) one sided p-values

*,** Indicate unit root at I(1) and I(2) respectively

Source: The Renewable Energy Index result from own calculations. The World Alternative Energy Index (WAEX), the Morgan Stanley Capital Index World Index (Stocks), the J.P. Morgan Global Government Bond index (Bonds), the C.S/Tremont Hedge Fund Index (hedge funds), the Liquid Private equity Fund 50 (Private Equity), the World Real Estate Investment Trust Index (Real Estate), the Dow Jones Emerging Market Index (Emerging Markets) and the Goldman Sachs Commodity Index (Commodities) are obtained from Thomson Financial Datastream. The Three Month U.S. T-Bill is obtained from St Louis FRED.

Johansen Co-integration Analysis

Table 6 exhibits the Johansen co-integration test results. The tests for co-integration are performed for those series whose variables were found to be non-stationary in both levels. Co-integration tests are meant to indicate whether there is a long term relation between renewable energy and cleantech and another particular series. Two null hypotheses are stated. First, there is no co-integrating relationship and second, there is at most one co-integrating vector, where the number of co-integrating vectors is pointed out by r. The number of lags refers to the number of lags used for the V.A.R model that is estimated for the purpose of the co-integration computation. Based on the eigenvalue and Trace test are number of lags chosen. The optimal outcome is presented in Table 6. No series holds a significant probability. Therefore, the null hypothesis cannot be rejected. The main finding, obtained from the Johansen co-integration tests, is that no long term relationships exist between renewable energy and the other asset classes studied. The same accounts for the cleantech sector. This implies that in the long run the prices for the various series may diverge. This also means that possible short term influences are not affected. When no long term relations were found, short term relations are sought after by performing Granger causality tests.

Granger Causality

Granger causality tests are performed on the basis of the unrestricted V.A.R. models described in the methodology section. The results are presented in Table 7. The overall weak results can perhaps be explained by the argument given previously; the renewable and cleantech sector are currently booming and are subject to influences very typical for their markets. As a consequence, short term causality is lacking in its presence for almost all relations. However, for renewable energy several short-term causal linkages are found. There is feedback between the real estate market and the renewable energy sector. Real estate is highly sensitive to certain economic states and it is a market where changes in the economic climate are felt first. These changes may be caused by the same factors that influence renewable energy. Thereby, real estate could function as a leading indicator of the renewable energy market.

Table 6
Johansen Co-integration Test Results

Table 6: exhibits the results for the Johansen co-integration test. Based on the unit root test the Three Month U.S T-Bill and Private Equity were excluded. Trend assumption: Linear deterministic trend - with intercept and no trend in CE and test VAR. Variables are divided into several categories. Data ranges from 1994 to 2008 ytd. Overall results are weak, no variables are significant. See the Correlation Matrix

Panel A - Renewable Energy							
Asset	H_0	H_1	Lags	Eigenvalue	Trace Statistic	5% Critical Value	Prob.[α]
Cleantech	r = 0	r > 0	4	0.0213	12.3419	15.4947	0.1413
	r ≤ 1	r > 1		0.0076	3.2211	3.8415	0.0727
Market							
Stocks	r = 0	r > 0	3	0.0167	8.3241	15.4947	0.4314
	r ≤ 1	r > 1		0.0028	1.1986	3.8415	0.2736
Corporate Bonds	r = 0	r > 0	1	0.0181	3.0532	15.4947	0.9431
	r ≤ 1	r > 1		0.0033	0.5443	3.8415	0.4606
Government Bonds	r = 0	r > 0	1	0.0197	3.3267	15.4947	0.9226
	r ≤ 1	r > 1		0.0007	0.1174	3.8415	0.7318
Government 30Y Bonds	r = 0	r > 0	1	0.0071	5.4054	15.4947	0.7643
	r ≤ 1	r > 1		0.0055	2.3691	3.8415	0.1238
Alternatives							
Hedge Funds	r = 0	r > 0	4	0.0159	4.8607	15.4947	0.8234
	r ≤ 1	r > 1		0.0012	0.3474	3.8415	0.5556
Real Estate	r = 0	r > 0	4	0.0097	5.6615	15.4947	0.7351
	r ≤ 1	r > 1		0.0037	1.5645	3.8415	0.2110
Emerging Markets	r = 0	r > 0	2	0.0161	8.1735	15.4947	0.4470
	r ≤ 1	r > 1		0.0031	1.3095	3.8415	0.2525
Commodities	r = 0	r > 0	2	0.0173	8.4490	15.4947	0.4187
	r ≤ 1	r > 1		0.0026	1.0834	3.8415	0.2979
Panel B - Cleantech							
Asset	H_0	H_1	Lags	Eigenvalue	Trace Statistic	5% Critical Value	Prob.[α]
Renewable Energy Index	r = 0	r > 0	4	0.0279	5.5287	15.4947	0.7504
	r ≤ 1	r > 1		0.0054	0.8846	3.8415	0.3469
Market							
Stocks	r = 0	r > 0	3	0.0554	9.8186	15.4947	0.2949
	r ≤ 1	r > 1		0.0018	0.3053	3.8415	0.5806
Corporate Bonds	r = 0	r > 0	1	0.0181	3.5975	15.4947	0.9333
	r ≤ 1	r > 1		0.0032	0.5443	3.8415	0.4606
Government Bonds	r = 0	r > 0	1	0.0197	3.4441	15.4947	0.9432
	r ≤ 1	r > 1		0.0007	0.1174	3.8415	0.7318
Government 30Y Bonds	r = 0	r > 0	1	0.0246	4.1619	15.4947	0.8900
	r ≤ 1	r > 1		0.0001	0.0208	3.8415	0.8851
Alternatives							
Hedge Funds	r = 0	r > 0	4	0.0336	9.3874	15.4947	0.3308
	r ≤ 1	r > 1		0.0139	2,7320	3.8415	0.0984
Real Estate	r = 0	r > 0	4	0.0232	3.9692	15.4947	0.9060
	r ≤ 1	r > 1		0.0002	0.0456	3.8415	0.8308
Emerging Markets	r = 0	r > 0	2	0.0175	3.0831	15.4947	0.9630
	r ≤ 1	r > 1		0.0011	0.1846	3.8415	0.6674
Commodities	r = 0	r > 0	2	0.0269	4.4806	15.4947	0.8612
	r ≤ 1	r > 1		0.0000	0.0056	3.8415	0.9394

* Indicates statistical significance at the 5% level. Probabilities are obtained from the Eviews 6.0 program
[α] Mackinnon-Haug-Michelis (1999) p-values

Source: The Renewable Energy Index result from own calculations. The World Alternative Energy Index (WAEX), the Morgan Stanley Capital Index World Index (Stocks), the J.P. Morgan Global Government Bond index (Bonds), the C.S/Tremont Hedge Fund Index (hedge funds), the Liquid Private equity Fund 50 (Private Equity), the World Real Estate Investment Trust Index (Real Estate), the Dow Jones Emerging Market Index (Emerging Markets) and the Goldman Sachs Commodity Index (Commodities) are obtained from Thomson Financial Datastream. The Three Month U.S. T-Bill is obtained from St Louis FRED.

Table 7
Granger Causality Test Results

Table 7: exhibits the Granger (short term) causality results for the Renewable Energy Index and the Cleantech sector. The same variables have been used as in the co-integration tests, using the same amount of lags as well. The null and alternative hypotheses are for example the Renewable Energy Index, H0: RE Index is not Granger caused by 'asset' and 'asset' is not Granger caused by RE Index and H1: RE Index is caused by 'asset' and 'asset' is caused by RE Index. # stands for 'not Granger caused by'. For the Renewable Energy Index 3 causal relations are found. In case of Cleantech one relation is found. See the Correlation Matrix, Table 3, for more information.

Panel A - Renewable Energy								
Asset			F-statistic					F-statistic
Cleantech	#	RE Index	0.94022	RE Index	#	Cleantech		143.634
			0.4423					0.2245
Market								
Stocks	#	RE Index	0.59314	RE Index	#	Stocks		179.239
			0.6681					0.1330
Corporate Bonds	#	RE Index	126875	RE Index	#	Corporate Bonds		185.072
			0.2846					0.1219
Government Bonds	#	RE Index	119.584	RE Index	#	Government Bonds		199.280
			0.3149					0.0983
Government 30Y Bonds	#	RE Index	232.101	RE Index	#	Government 30Y Bonds		284.387
			0.0593					*0.0260
Alternatives								
Hedge Funds	#	RE Index	141.614	RE Index	#	Hedge Funds		211.140
			0.2311					0.0819
Real Estate	#	RE Index	251.478	RE Index	#	Real Estate		0.75559
			*0.0438					0.5557
Emerging Markets	#	RE Index	0.83971	RE Index	#	Emerging Markets		0.83971
			0.5019					*0.0448
Commodities	#	RE Index	136.387	RE Index	#	Commodities		189.417
			0.2490					0.1141
Panel B - Cleantech								
Asset			F-statistic					F-statistic
Market								
Stocks	#	Cleantech	152.878	Cleantech	#	Stocks		196.761
			0.1964					0.1021
Corporate Bonds	#	Cleantech	170.053	Cleantech	#	Corporate Bonds		191.089
			0.1526					0.1113
Government Bonds	#	Cleantech	123.928	Cleantech	#	Government Bonds		197.959
			0.2965					0.1003
Government 30Y Bonds	#	Cleantech	228.220	Cleantech	#	Government 30Y Bonds		323.935
			0.0630					*0.0139
Alternatives								
Hedge Funds	#	Cleantech	155.818	Cleantech	#	Hedge Funds		115.331
			0.1882					0.3337
Real Estate	#	Cleantech	156.530	Cleantech	#	Real Estate		133.958
			0.1862					0.2577
Emerging Markets	#	Cleantech	0.87565	Cleantech	#	Emerging Markets		208.503
			0.4800					0.0853
Commodities	#	Cleantech	125589	Cleantech	#	Commodities		131.609
			0.2898					0.2664

* Indicates statistical significance at the 5% level. Probabilities are obtained from the Eviews 6.0 program

Probabilities are Mackinnon-Haug-Michelis (1999) p-values and displayed in italics

Source: The Renewable Energy Index result from own calculations. The World Alternative Energy Index (WAEX), the Morgan Stanley Capital Index World Index (Stocks), the J.P. Morgan Global Government Bond index (Bonds), the C.S/Tremont Hedge Fund Index (hedge funds), the World Real Estate Investment Trust Index (Real Estate), the Dow Jones Emerging Market Index (Emerging Markets) and the Goldman Sachs Commodity Index (Commodities) are obtained from Thomson Financial Datastream.

The second granger cause is renewable energy appearing to influence the emerging markets in the short term. Already, renewable energy showed strong correlation with emerging markets in the correlation matrix in Table 3. Maybe renewable energy is more sensitive to economic circumstances and emerging markets, primarily driven by their strong organic growth, less receptive to economic changes, which causes emerging market prices to lag behind price changes in renewable energy. A third Granger cause, why the renewable energy sector causes 30-year government bonds, is more difficult to reason.

For cleantech the Granger causality tests are specifically weak, only one short term relation is found. The single null hypothesis that can be rejected is cleantech is not Granger caused by 30 year government bonds. This also counts for renewable energy, but a reasonable economic explanation is however absent. These poor results and lack of economic viability is the reason why this section is moved to the appendix. Keeping it in the empirical section would only distract attention from the subject covered and mitigate the power of the message which this thesis aims to tell. The Granger causality tests were the last tests of the empirical analysis.

References

Ackermann, C., R. Mcenally, D. Ravenscraft, 1999. The Performance of Hedge Funds: Risk, Return, and Incentives. *The Journal of Finance* 3, June 1999: 833-874.

Alexander, C., I. Giblin, and W. Weddington III., 2001. Co-integration and Asset Allocation; A New Active Hedge Fund Strategy. *Discussion Papers in Finance ISMA Center.* 2001-2003, University of Reading.

Ang, A., Chen, J. and Y. Xing, 2006. Downside Risk. *Review of Financial Studies* 19: 1191-1236.

Anon. *Our common future World Commission on Environment and Development*, Oxford University Press, Oxford (1987).

Bawa, V.S. and E.B. Lindenberg, 1977. Capital Market Equilibrium in a Mean-Lower Partial Moment Framework. *Journal of Financial Economics* 12: 129-156.

Bikker, J., D. Broeders and J. de Dreu, 2007. Stock market performance and pension fund investment policy: rebalancing, free float, or market timing? *DNB Working Papers 154*, Netherlands Central Bank, Research Department.

Brennan, M.J., E.S. Schwartz an R. Lagnado, 1997. Strategic asset Allocation, *Journal of Economic Dynamics and Control* 21: 1377-1403.

Brown M.A., M. D. Levine, W. Short and J. G. Koomey, 2001. Scenarios for a Clean Energy Future, *Energy Policy* 29 (2001) 1179–1196

Cambell, J.Y. and L.M. Viceira, 2000. *Strategic Asset Allocation - Portfolio Choice for Long Term Investors* . Published: Oxford University Press, 2000.

Cheung, Y.W. and K.S. Lai, 1993. Finite-sample Sizes of Johansen's Likelihood Ratio Tests for Cointegration. *Oxford Bulletin of Economics and Statistic* 55: 313-328.

Chien, T. and J-L. Hub, 2008. Renewable energy: An efficient mechanism to improve GDP, *Energy Policy* 33 (2008): 3045-3052

Dincer, I , 2003. Renewable energy and sustainable development: a crucial review, *Renewable and sustainable energy review* 4 (2003): 157-175

Elliott, D., 2000. Renewable Energy and Sustainable Futures, *Futures* 32: 261-274

Energy Information Administration (EIA), 2005. *Policies to Promote Non-hydro Renewable Energy in the United States and Selected Countries*, February 2005. United States Department of Energy, Washington, DC 20585.

Energy Information Administration (EIA), 2005. *Annual energy outlook*. United States. http://www.eia.doe.gov/oiaf/archive/aeo05/index.html

Energy Information Administration (EIA), 2005. *Annual Energy Outlook 2008, With Projections to 2030*. United States Department of Energy, Washington, DC 20585.

Engle, R.F. and C.W.J. Granger, 1987. Co-integration and Error Correction Representation, Estimation, and Testing. *Econometrica* 55: 251-276.

Estrada, J., 2002. Systematic Risk in Emerging Markets: the D-CAPM. *Emerging Markets Review* 3 (2002): 365-379.

Fama, E. and K. French, 2003. The CAPM: Theory and Evidence, *Center for Research in Security Prices*, Working paper 550 (2003)

Faulina, J., F. Lerab, J.M. Pintorc and J. Garcia, 2006. The Outlook for Renewable Energy in Navarre: An Economic Profile, *Energy Policy* 34 (2006): 2201–2216

Gompers, P. and J. Lerner, 2000. Money chasing deals? The impact of fund inflows on private equity valuation, *Journal of Financial Economics* 55: 281–325.

Haas, R. et al., 2004. How to Promote Renewable Energy Systems Successfully and Effectively, *Energy Policy* 32: 833-839

Hamilton, S., H. Jo and M. Statman, 1993. Doing well while doing good? The Investment Performance of Socially Responsible Mutual Funds, *Financial Analysts Journal* 49: Nov/Dec 1993, 6: 62-68

Hartley, D.L., 1990. Perspectives on renewable energy and the environment, *Energy and the environment in the 21st Century*. Massachusetts: MIT, 1990.

Hoevenaars, R.P.M.M., R. Molenaar, P. Schotman, T.B.M. Steenkamp, 2007. Strategic Asset Allocation with liabilities: Beyond Stocks and Bonds. *Journal of Economic Dynamics and Control* 32, Issue 9, September 2008: 2939-2970

Lintner, J., 1965. The Valuation of Risky Assets and the Selection of Risky Investments in Stock portfolios and Capital Budgets. *Review of Economics and Statistics* 47: 13 - 37.

Lee, S. and S. Stevenson, 2003. Time Weighted Portfolio Optimisation, *Journal of Property Investment and Finance* 21, No. 3: 233-249.

Klaas, R., M. Goovaerts, J. Dahne and M. Denuit, 2004. *Modern Actuarial Risk Theory*. Published: Springer, 2004.

Margolis, R. M. and D. M. Kammen, 1999. Evidence of Under-Investment in Energy R&D in the United States and the Impact of Federal Policy, *Energy Policy* 27: 575-584.

Markowitz, H.M., 1952. Portfolio Selection. *Journal of Finance* 7: 77-91.

Nawrocki, D., 1999. A Brief History of Downside Risk Measures." *Journal of Investing* (Fall 1999): 9-25

Ogden, J.M., 1999. Prospects for building a hydrogen energy infrastructure, *Annual Review Energy Environment* 24: 227–79

Painuly J. P., Barriers to renewable energy penetration; a framework for analysis, *Renewable Energy* 24, Issue 1, Sept. 2001: 73-89

Phillips, P.C.B., 1991. Optimal Inference in Cointegrated Systems. *Econometrica* 59: 283-306.

Post, T. and P. van Vliet, 2004. Conditional Downside Risk and the CAPM, *Erasmus School of Economics*, working paper (2004).

Ray, K. and D. Nawrocki, 1996. Linear Adaptive Weights and Portfolio Optimization, available at: www.handholders.com/old/raylam.html

REN21, Renewables Global Status Report 2007, Worldwatch Institute, available at: http://www.ren21.net/pdf/RE2007_Global_Status_Report.pdf

Sawin, J.L., 2001. The Role of Government in the Development and Diffusion of Renewable Energy Technologies: Windpower in the United States, California, Denmark and Germany, 1970-2000, PhD Dissertation, Fletcher School of Law and Diplomacy, Tufts University, September 2001, Chapter 5.

Sharpe, W.F., 1964.Capital Asset Prices: A Theory of Market Equilibrium under Conditions of Risk, *Journal of Finance* 19 (1964): 425-442.

Sharpe, W.F., 1994. The Sharpe Ratio. *The Journal of Portfolio Management*. Fall 1994

Sharpe, W.F., 1995. The Style and Performance of Large Seasoned US Mutual Funds, 1985-1994", available at www.stanford.edu/~wfsharpe/art/ls100/ls100.html, March.

Sorenson, B.E., 2005. Granger Causality. *Economics 7395*. Spring, 2005.

Wustenhagen, R. and M. Bilharz, 2006. Green energy market development in Germany: effective public policy and emerging customer demand. *Energy Policy* 34 (2006): 1681-1696.

Wong, A.S.K, and P.J.G.Vlaar, 2003. Modelling Time-Varying Correlations of Financial Markets, *Econometric research and special studies department, Research memorandum* No. 739. De Nederlandsche Bank.